Disarming Doomsday

Radical Geography

Series Editors:
Danny Dorling, Matthew T. Huber and Jenny Pickerill
Former editor: Kate Derickson

Also available:

Unlocking Sustainable Cities:
A Manifesto for Real Change
Paul Chatterton

In Their Place:
The Imagined Geographies of Poverty
Stephen Crossley

Making Workers:
Radical Geographies of Education
Katharyne Mitchell

Space Invaders:
Radical Geographies of Protest
Paul Routledge

New Borders:
Migration, Hotspots and
the European Superstate
Antonis Vradis, Evie Papada,
Joe Painter and Anna Papoutsi

Disarming Doomsday

The Human Impact of Nuclear Weapons since Hiroshima

Becky Alexis-Martin

First published 2019 by Pluto Press
345 Archway Road, London N6 5AA

www.plutobooks.com

Copyright © Becky Alexis-Martin 2019

The right of Becky Alexis-Martin to be identified as the author of this work has been asserted by her in accordance with the Copyright, Designs and Patents Act 1988.

British Library Cataloguing in Publication Data
A catalogue record for this book is available from the British Library

ISBN   978 0 7453 3921 4   Hardback
ISBN   978 0 7453 3920 7   Paperback
ISBN   978 1 7868 0438 9   PDF eBook
ISBN   978 1 7868 0440 2   Kindle eBook
ISBN   978 1 7868 0439 6   EPUB eBook

This book is printed on paper suitable for recycling and made from fully managed and sustained forest sources. Logging, pulping and manufacturing processes are expected to conform to the environmental standards of the country of origin.

Typeset by Stanford DTP Services, Northampton, England

Simultaneously printed in the United Kingdom and United States of America

Contents

List of Figures and Tables vi
Series Preface vii
Acknowledgements viii

1 The Radical Geography of Nuclear Warfare 1
2 A Secret History 9
3 The Mystery of the X-ray Hands 30
4 After Nuclear Imperialism 53
5 After Nuclear War 68
6 Strange Cartographies and War Games 79
7 Spaces of Irregularity 91
8 Spaces of Peace 106
9 Future War Zones 125

Notes 139
Index 166

List of Figures and Tables

FIGURES

2.1	US military damage maps produced in 1946 demonstrate the extent of damage to Hiroshima and Nagasaki	21
2.2	Labourers working on the restoration of Hiroshima's Aioi Bridge in 1949	23
2.3	Los Alamos, a place of discoveries and secrets	24
2.4	Public incentives not to investigate the A-bomb dome too closely	28
2.5	Some of the Hiroshima Society of Patriots protesting outside the Peace Park on Peace Day, 6 August. They did not like me as much as this photo suggests	29
3.1	Young soldiers hanging out on Christmas Island	40
3.2	The Chairmen of the French and British atomic veterans' associations rekindling the flame under the Arc de Triomphe, Paris	51
5.1	The current public geography of LANL reveals little about this place	73

TABLES

3.1	The UK nuclear tests undertaken during the atmospheric testing series, including numbers of participants for each operation	34
7.1	Nuclear defence treaties: past and present (nuclear-free treaties highlighted in bold)	95

Series Preface

The Radical Geography series consists of accessible books which use geographical perspectives to understand issues of social and political concern. These short books include critiques of existing government policies and alternatives to staid ways of thinking about our societies. They feature stories of radical social and political activism, guides to achieving change, and arguments about why we need to think differently on many contemporary issues if we are to live better together on this planet.

A geographical perspective involves seeing the connections within and between places, as well as considering the role of space and scale to develop a new and better understanding of current problems. Written largely by academic geographers, books in the series deliberately target issues of political, environmental and social concern. The series showcases clear explications of geographical approaches to social problems, and it has a particular interest in action currently being undertaken to achieve positive change that is radical, achievable, real and relevant.

The target audience ranges from undergraduates to experienced scholars, as well as from activists to conventional policy makers, but these books are also for people interested in the world who do not already have a radical outlook and who want to be engaged and informed by a short, well written and thought-provoking book.

Danny Dorling, Matthew T. Huber and Jenny Pickerill
Series Editors

Acknowledgements

This book is dedicated to my PhD supervisor David Martin. Thank you for everything.

I have deployed
the nukes
that were in
the stockpile

and which
you were probably
saving
for arms control.

Forgive me
I was capricious
So rash
and so bold.

(Becky Alexis-Martin after William Carlos Williams, 2018)

1
The Radical Geography of Nuclear Warfare

> What the map cuts up, the story cuts across.
> – Michel de Certeau[1]

Nuclear warfare is geographically distinct from other forms of warfare. It can shrink space and time to cause rapid mass destruction, reducing dependence on the operational military geographies and planned time accounting for distance and terrain.[2] Nuclear warfare is also a phenomenon of blurred boundaries. It is everywhere and nowhere, it transcends war and peace, it amalgamates war zone and homeland, and it merges the real and virtual.[3,4]

Nuclear warfare is bounded by geography, there are multiple elements that make up the assemblages of nuclear war, consisting of many different landscapes, geotechnologies, zones, bodies and communities.[5] The processes that surround the nuclear military industrial establishment, from nuclear weapon design to deployment, are inherently geographical in nature. The geographies of nuclear warfare include the geopolitics of nuclear strategy, and explore the fractured internationalism of the Global North–South nuclear divide. The geographical nature of irregular nuclear warfare and state-sanctioned terrorism is explored; as is the significance of geotechnologies, including Geographical Information Systems (GIS) and Remote Sensing. The material artefacts of the nuclear military industrial complex, from bunkers to bombs, are also relevant, as is their relationship to each other according to theories of assemblage. Also significant are the places and spaces that have been left unmapped or are unmappable due to the culture of secrecy that surrounds the military industrial complex.

Understanding the radical geographies of nuclear warfare should begin with the consideration of what is left unmapped. Cartography can be an elegant resource for understanding the world but is still limited by what is shown or concealed by the map maker. Maps can be sources of epistemic injustice. Maps have been used to deny knowledge, and as a tool to disguise the presence of the secret places where nuclear weapons were developed. They can be used to hide the traces of things that history would rather forget, a fresh green veneer for the jagged surfaces of military nuclear accident. They camouflage things that were perhaps never mapped to begin with, emerging in a flurry of pollution only to subsume into re-wilded nature reserve. Once it has gone, even the ground-truthing of satellite imagery may not reveal the traces of what came before. The perpetual remapping and remaking of nuclear warfare both constitutes and is reliant upon a concerning institutional and cultural amnesia.

The birth of nuclear warfare heralded the coming of the secret city. Entire settlements came into being beyond the public domain, hidden from and by cartographers in the USA, the UK, France, China and the USSR, their sole purpose being the covert advancement of nuclear warfare technologies. For example, in 1943 an isolated nuclear defence research laboratory was established in Los Alamos, New Mexico. The 'Plutopia' of Los Alamos was so clandestine that initially it didn't even have a name; it was referred to instead as 'Project Y'.[6] The unmapped streets of Los Alamos had the dual benefits of providing national security by concealment, and of isolating participants from a potentially disapproving outside world. The town was not included on any formal documentation or postal mail, and the whole area was fenced off and surrounded by guards. Academics were intellectually imprisoned by the legislation that entwined the nuclear military industrial establishment, silenced by Espionage Acts and Military Secrets Acts. For scientists during the austere years of the Second World War, this echoes the discourse of Yevgeny Zamyatin's dystopian novel *We*, 'Those in paradise were given a choice: happiness without freedom, or freedom without happiness. There was no third alternative.'[7] Los Alamos was designed to embody the respectable atmosphere of a campus town environment, with careful planning to demilitarise the feel of the place. Picket fences and leafy avenues created a sense of normality that encouraged scientists and their families to relocate. Excellent health care, good schools, churches and leafy parks simultaneously incentivised, suburbanised and banalised the work of scientists who were employed to create weapons of mass destruction.

Los Alamos did not occur in isolation; this was an international phenomenon. In 1958, the Chinese settlement No. 404 Factory of China

National Nuclear Corporation was covertly built to refine plutonium and produce the components for nuclear weapons. The greatest number of secret cities emerged across what was then the USSR.[8] Similarly to those living in Los Alamos, residents of the USSR secret cities received privileges, such as better food, health care and quality of life. However, their freedoms were severely restricted. Secret cities have created long-term challenges for former Soviet bloc countries due to the consequences of delayed under-urbanisation. This is the way that spatial patterns of population change have been shaped by closed cities, creating population loss, socio-spatial polarisation, and the re-emergence of social-spatial patterns that are attributed to under-urbanisation. This reflects the limited economic resources that are available to rural communities surrounding once-secret cities, despite the abundant opportunities provided to city residents.[9] It also demonstrates some of the subtler sociocultural consequences of nuclear secrecy. It is not yet possible to universalise the consequences of opening secret facilities, as this is not merely a historic phenomenon. There are still atomic cities that remain closed to prying eyes. An estimated 1.5 million people still live in the publicly acknowledged ones in Russia, with yet more unknown nuclear enclaves potentially hidden from view.[10] In the USA, Mercury, Nevada, was used for nuclear weapons testing from 1951 to 1992, and remains closed to the public to this day.[11] Its population has decreased from its Cold War heyday, but 500 people still live in Mercury. Only with time, and perhaps with the work of some very careful radical spatial analysists, will more of the mysteries of secret cities and other hidden nuclear spaces be revealed.

It is not just maps that obfuscate. Military euphemisms have long been a tactic for the concealment of inconvenient truths, designed to limit our understanding of the global harm caused by the nuclear military industrial establishment. Every thermonuclear bomb has a legacy of social and environmental harm and could obliterate the lives of thousands of people if deployed. However, nuclear weapon possessor states often attempt to anaesthetise the impact of this. In their sanitised language, arsenals of nuclear weapons are a single 'nuclear deterrent' constructed from 'ICBMs' (intercontinental ballistic missiles) rather than international deadly warhead flingers. Whether the purpose of this deterrence is to deter unwelcome visitors, or merely other nuclear weapon possessor states is not explicitly stated. The five Non-Proliferation Treaty (NPT) designated nuclear weapon possessor states of the USA, UK, France,

Russia and China are among the most powerful in the world. They possess other equally powerful forms of conventional warfare, such as the American MOAB (Mother of All Bombs or Massive Ordinance Air Blast, depending on your perspective), which has the killing capacity of a nuclear weapon and was deployed in Afghanistan by President Trump in 2017. Nuclear warfare is somehow made respectable by omission, while the production, possession and detonation of nuclear weapons remains an abnormally murderous blip in human history. A culture of geopolitical discussion in abstract rather than concrete terms has meant that the Cold War was the first war to begin without a declaration of war and to end without a peace treaty.

Multilateral nuclear disarmament is eminently preferable to the current volatile nuclear imperialism and necropolitics of nuclear weapon possessor states. It is culturally, socially, economically and environmentally astute to disarm. It would release funds for more purposeful and altruistic aims, such as health care and development. It would sever the chain of racism, corruption and exploitation – from uranium mining, to the development and testing of new nuclear weapons. It is outrageous that the slow violence of nuclear warfare is still even considered to be a worthy project for our economic capital and our brightest minds. Nuclear development wastes such minds by stifling their capacity for critical thinking. It wastes them by channelling their purpose and brilliance into one function, specifically creating and maintaining the conditions to sustain the perfect machine of mass-murder. The nuclear realm is reliant upon the anthems, flags, medals and uniforms of the military establishment to maintain its respectability in the public domain.

This is an era of precarious peace. While the majority of states in the nuclear-free Global South have cooperated with the United Nations (UN) to ban nuclear warfare, nuclear weapon possessor states are still ignoring and boycotting the Treaty on the Prohibition of Nuclear Weapons:[12] 90 per cent of nuclear weapons are owned by rivals Russia and the USA, and there are still enough weapons to end life on this planet. However, taking a radical geographical approach to anti-nuclear pacifist action can disrupt these power imbalances by revealing the widespread and diverse environmental and social injustice of nuclear warfare. It can subvert and positively transform the mendacious language, spurious maps and models; unreliable histories; and the unacceptable geopolitical, cultural and social ideas that have been perpetuated by purveyors of nuclear war.

WHY IS RADICAL GEOGRAPHY NEEDED?

A radical geography of nuclear warfare aims to understand the social and spatial problems created by nuclear weapons, and to advocate solutions. The radical approach is committed to intellectualising, supporting and building alternatives to nuclear warfare, and is well-placed to create change due to its position between critical theory and activist praxis.

Radical geography is postcolonial. It is against the ecological imperialism of nuclear weapons; the cultural, social and ecological terrorism of colonisation. It is focused upon understanding the alterity of previously colonised peoples. For instance, it considers the consequences of nuclear weapons to the lives of the people who lived on the Pacific Islands during the time of the British, French and American nuclear weapons tests. It considers the experience of the subaltern, who may still be subject to the heavy-handed hegemony of nuclear imperialism. For example, the experiences of historic uranium miners in Africa and North America, who have been subjugated and exploited by mine owners.

It is feminist, acknowledging the significance of ecofeminism, recognising the power imbalances and linkages between the domination of women in patriarchal society and the subjugation of nature by the nuclear military industrial complex.[13] Patriarchy exemplifies a society where men are in control of the principal organs of power. Consider the volatile and hostile interactions between President Donald Trump and Supreme Leader Kim Jong-un, and if these interactions are politically healthy and ethically astute. Men are predominantly the leaders and perpetuators of our nuclear weapon possessor states. The type of power that these leaders engage with is inherently patriarchal and hierarchical, and it needs to be dismantled.

One notable exception to this discourse is UK Conservative Prime Minister Theresa May, who has declared that she would deploy nuclear weapons and kill hundreds of thousands of people, should the opportunity arise. She said 'yes, I would push the nuclear button'.[14] May is the contemporary of Labour leader and peace award-winner Jeremy Corbyn, who has expressed a commitment to nuclear disarmament that has been ridiculed as a retro hard-left fantasy by many in the UK. This is despite his pledge to increase nuclear-powered submarines instead of those armed with nuclear weapons, to support the industry, and despite the majority of countries in the world agreeing with his aspirations.[15]

Theresa May also follows in the footsteps of Conservative Prime Minister Margaret Thatcher, who did much to support Reagan and Gorbachev's negotiations to end the Cold War and to begin denuclearisation. At the 1986 Reykjavik summit, Reagan and Gorbachev came surprisingly close to eliminating much of their nuclear arsenals. They also brought about the Intermediate-range Nuclear Forces Treaty, which banned short- and medium-range nuclear missiles, reducing the risk of nuclear warfare.[16] President Trump now says he plans to withdraw from this landmark treaty. So, where did we go wrong? Changing the acceptable status of nuclear warfare will challenge the patriarchy, and build space for better pacifist, feminist and human-security based approaches to politics. Radical geography posits the possibility of resistive action as an option for everyone, as a source of pacifist acts of cultural and social defiance against nuclear weapons.

If you do stand for radical geographical action against nuclear war, then you are in good company. There is an extensive history of pacifist non-violent action against nuclear weapons. The acts of occupation and protest are inherently spatial in nature, and often include marginalised communities, from the women of Greenham Common, the creators of the world's longest-lasting peace camp, to Albert Einstein and George Orwell, the radical intellectuals who decided to take a stand against nuclear war. And from the astonishingly brave civil disobedience of the Committee of 100 in the 1960s to the smart commentary provided by Beatrice Fihn of the International Campaign to Abolish Nuclear Weapons (ICAN), which has helped to lay the groundwork for an international UN Nuclear Ban Treaty in 2017.

There are many geographers who have explored the issues surrounding nuclear defence with a critical gaze. Notably, a Cold War rally for pacifist geographies of nuclear war was undertaken by Pepper and Jenkins in 1983, who had answered their own call with *The geography of peace and war* by 1985.[17] This anthology included essays by qualitative and quantitative human geographers on topics such as teaching peace, Soviet pacifist nuclear geographies, and doomsday scenarios. Their work has been followed by that of sympathetic ethnographers, geographers and social scientists, who have all engaged critically with the issue of nuclear defence. It is notable that many of these scholars are women. For instance, Kate Brown undertook a deep historic ethnography of two communities of nuclear defence workers in the USA and the former Soviet Union. Her work in *Plutopia: Nuclear families, atomic cities and the great Soviet and American plutonium disasters* reveals some of the secrets of the cities of Richland and Ozersk, and shows how Cold War attitudes have remained pervasive long after the lifting of the Iron Curtain.[18] Stephanie Malin has worked closely with uranium mining communities and the nuclear site down-winders of the USA as both an academic and a social justice activist. Her work in *The price of nuclear power: Uranium communities and environmental justice* describes the economic precarity and exploitation of indigenous and local uranium mining communities, exploring their unique battles for recognition and justice.[19] The uranium mine features as a space of contestation and occupation by the military industrial establishment and state. Another noteworthy accidental radical geographer is Kristen Iversen, author of *Full body burden: Growing up in the nuclear shadow of Rocky Flats*.[20] Her writing juxtaposes her personal experiences with the local culture of nuclear manufacturing, providing a very personal and beautifully written retrospective psychogeography of Rocky Flats plutonium processing plant. These are just some of my personal favourites, there are many others who also apply a critical and spatial gaze to their work on the processes of nuclear warfare.

There are radical geographers who are occupying spaces, undertaking protests and using pacifist non-violent action against nuclear war, right now. Among them are the British anarchist geographers Kelvin Mason, Phil Johnstone and Kye Askins. For these academics, the role of the academic is visibly and publicly merged with that of sympathiser and of social justice activist. They have undertaken academic activism in the spaces that surround the Atomic Weapons Establishment for several years, in order to explore the existential ghosts of place and harmful activity that surround British nuclear weapon manufacturing.[21] In their most recent project, entitled 'Nuclear Refrain', this dynamic collective states that the purpose of their crowd-funded work is to critique the UK's Trident nuclear defence system, and to 'transgress writing genres to disrupt the usual nuclear weapons narratives, seeking to unsettle and so expose considerable alternatives'.[22] As they disrupt and unsettle the entrances of nuclear defence sites, they demonstrate that radical geographical action can go beyond pataphysics to provide an effective riposte to the latent threat of nuclear weapons. Unlike other major global challenges, such as climate change, there is no need to reach for a new science of imaginary solutions. Disarmament technologies are already demonstrably effective. Disarmament policy has already created a nuclear-free Global South. Nuclear warfare is not politically unconscious.

A liminal state of nuclear warfare persists, since weapons persist, and remain harmful, despite radical progress in international policy. There is still a repressive tolerance towards nuclear weapons. This manifests as unthinking acceptance of their necessity, despite their obvious capacity for harm to both humans and the environment. It includes the vocal endorsement of actions that are manifestly aggressive towards humans, such as the bombastic rhetoric of President Donald Trump in 2017 towards the leader, and therefore the people, of the Democratic Republic of North Korea. Trump's warmongering and flawed synecdoche does not represent the perspective of most Americans.

Fresh challenges have recently arisen as aspiring nuclear states, such as Iran and North Korea, vie to engage in nuclear proliferation. While economic sanctions and international legislation used to be effective during the Cold War, they no longer work. The barriers to entry are now lower. Most states would have the capacity to harness now-archaic nuclear technology, and it is no longer as challenging for aspiring nuclear states to develop native programmes. Threatening nuclear war with states that attempt to proliferate is uncivilised, undignified and no longer enough to prevent proliferation.

There are currently five nuclear weapon states under the terms of the NPT, namely the USA, Russia, France, China and the UK. A further three states, India, Pakistan and North Korea, are known to have nuclear weapons. Israel is also a possessor of nuclear weapons, but is not yet classified as such due to military ambiguity. It is currently estimated by the *Bulletin of Atomic Scientists* that there are 9,220 nuclear weapons in the world. This is a notable reduction compared to the number of nuclear weapons during the Cold War era, but still 9,220 more than we need. To move forward from our current nuclear predicament, we need a situated understanding of nuclear warfare in space, place and time. We need to develop ways to explore the half-truths that surround the nuclear military industrial establishment. We need to establish what exactly nuclear weapons are supposed to deter and why. We need to create our own peace games, to counter the symbolic and actual violence of nuclear war.

2
A Secret History

> How could we have done this ultimately terrible and irresponsible thing? Sixty years is enough! Stop doing this immoral thing. Weapons scientists, REVOLT.
>
> – Ed Grothus, Preacher of the Critical Mass

The radical geographies of nuclear warfare have an inauspicious beginning, as secret 'closed' cities devoted to the nuclear weapons flourished across the world during the Cold War. Los Alamos is one of those cities that rose from obscurity, nestled in the canyons of New Mexico. It was to become infamous as the home of the atomic bomb. It was notable as the source of the science and technology that propelled and detonated atomic bombs over Hiroshima and Nagasaki on 6 and 9 August 1945. It sank out of prominence after this brutal event. For the *Hibakusha*, the people who were exposed to the blast and ionising radiation from nuclear warfare, there was little anonymity and their lives were forever changed. The bombing of Hiroshima and Nagasaki demonstrated that the US had developed a weapon that was bigger, stronger, brighter and more powerful than any other. The advent of the atomic age heralded a new era of nuclear risks, security challenges and humanitarian issues, and no one was well-equipped to deal with the consequences.

Now, nuclear warfare affects almost every person on Earth. It is not just the hundreds of surviving *Hibakusha* of Japan who are affected; or the hundreds and thousands of military atomic test veterans worldwide who unleashed increasingly powerful atom bombs, and then turned to larger and more devastating thermonuclear weapons or hydrogen bombs. It is not just the tens of thousands of people who have been displaced from their homes internationally across decades, becoming nuclear refugees, with their landscapes polluted by heavy radioisotopes. It is not just the thousands of workers and residents of nuclear weapon manufacturing towns, who face uncertainty around their health due to safety scares and leak cover-ups, who are affected.

It is everyone who pays taxes to a government that supports nuclear defence, and is unwittingly supporting nuclear warfare, to the detriment of public services and their own income. It is the vast swathe of people who already live in one of the nuclear-free zones of South America, the South Pacific, Africa or Asia, who do not want this polluting arsenal to unwittingly and uncontrollably cross their state boundaries. The secrecy that surrounds the development of the first atomic bomb and persists around nuclear weapons manufacturing can offer insights into why we still exist at a nuclear impasse.

The nuclear attacks on Hiroshima and Nagasaki are lauded for ending the Second World War. However, this is a naïve perspective with a complex history behind it, mired in American experimentation and domination. American President Franklin D. Roosevelt sanctioned the development of the world's first ever nuclear weapon in 1939, due to concerns about a new German atomic bomb.[1] Nuclear technology was a fiercely protected secret during the Second World War, so it is no coincidence that the USA was a pioneer, as its vast and unexamined spaces left it well-placed to develop nuclear science away from prying eyes.

In the 1940s, before the geographical insights of satellite imagery and high-flying spy planes were available, it was easier to obscure an entire city if a nation chose to do so. Los Alamos was pioneered as a purpose-built secret city for the design and manufacture nuclear weapons.[2] By 1942, the startling ideas emerging from this research laboratory had grown to become the Manhattan Project – an American, UK and Canadian nuclear defence research collaboration across several secret sites in the USA. Burgeoning places and spaces of nuclear defence began to spring up – from uranium mines to places for research, testing and production of the bomb. The Manhattan Project included the creation of Hanford nuclear laboratories, Washington; the Y-12 National Security Complex, in Oak Ridge, Tennessee; and Los Alamos, New Mexico. It rapidly grew to employ over 125,310 workers in the USA by 1944. This was the start of the *nuclear military industrial complex* – the complex alliance between the military and the nuclear weapons manufacturing sector, working together to influence public policy and international geopolitics.

Initially known as Project Y, Los Alamos was a secret city, invisible beyond national boundaries, unmapped and clandestine, yet home to thousands of scientists and their families.[3] To encourage scientists to relocate, the town was designed as a small-town campus environment for families rather than soldiers, with schools, churches, theatres and dance halls. Isolated and beautiful – officially, the campus of Los Alamos did not exist.[4] The town was not included on any formal documentation or postal mail, and the whole area was fenced off and surrounded by guards. A combination of academic and military leadership coalesced to create this town, and to solve the strategic and scientific challenges that surrounded the atomic bomb. By the end of the Second World War, this clandestine place had a hidden population of 6,000 and covered 43 square miles.

Led by the theoretical physicist, Dr J. Robert Oppenheimer, the scientists of Los Alamos had developed three useable weapons by mid-1945. The first nuclear weapon was detonated during the Trinity test at the Alamogordo range on 16 July 1945, with about a 20-kiloton yield.[5] There were concerns that the first nuclear weapon might ignite the global atmosphere and destroy the Earth, but this was seen as a trivial and unlikely risk, and testing was undertaken regardless. This is perhaps the first recorded instance of environmental concern relating to nuclear weapons. We now know that no detonation on Earth is ever likely to start an uncontrolled fusion reaction in the atmosphere. However, at the time, it was not made known that something could go wrong, because: 'There might be a reaction against science in general which would result in suppression of all scientific freedom and the destruction of science itself.'[6] It was a big risk for a relatively small gain.

The creation and testing of the Trinity bomb unleashed a rapid chain of events, leading to the bombings of Hiroshima and Nagasaki just one month later. The time from testing to deployment was startlingly short, and Germany had been defeated at this point, meaning that a Nazi atomic bomb was no longer a possibility. However, a Nazi nuclear weapon had never really been a plausible scenario, due to challenges surrounding scientific coordination and a lack of shared knowledge. Regardless, President Truman continued to pursue the use of atomic weapons in warfare, ordering the detonation of Little Boy over Hiroshima, on 6 August 1945. This was swiftly followed by the atomic bombardment of Nagasaki on 9 August with a different type of atomic bomb, to effectively create a human nuclear weapon testing site.[7] Nagasaki was the second nuclear weapon dropped in warfare, and there is hope that it may also be the last one.

The incremental atomic discoveries that led scientists at Los Alamos to this point were clinically documented by Henry DeWolf Smyth in a tome entitled *Atomic energy for military purposes: The official report on* the development of the atomic bomb under the auspices of the United States government, 1940–1945, published by Stanford University Press in 1945, after the bomb had been deployed in Japan. This enabled a few of the countless secrets of Los Alamos to be revealed.

HIROSHIMA AND NAGASAKI

On 6 August 1945, at quarter past eight in the morning, the B-29 bomber *Enola Gay* released Little Boy above Hiroshima.[8] This 16-kilotonne atomic bomb killed approximately 135,000 people, a third of the city's 420,000 residents. About 70,000 died instantly. No one knows the exact numbers, as many men had been conscripted during the Second World War, and the registry office was destroyed in the bombing. Hiroshima was selected as a target because it was home to a significant military base and had received no damage from prior air raids. A second bomb, Fat Man, was dropped on 9 August 1945. The bomb hit Nagasaki at quarter past nine in the morning and is estimated to have killed approximately 50,000 people in total, including 40,000 people who died instantly.[9] A third of Nagasaki was destroyed.

So, how did Nagasaki and Hiroshima become the chosen sites of atomic bombing by the USA? 'A list of suitable cities was curated, with a focus upon large urban areas of not less than 3 miles in diameter… between the Japanese cities of Tokyo and Nagasaki…and of high strategic value.'[10] The list included Hiroshima, Tokyo Bay, Kawasaki, Yokohama, Nagoya, Osaka, Kobe, Kyoto, Kure, Yawata, Kokura, Shimosenka, Yamaguchi, Kumamoto, Fukuoka, Nagasaki, Sasebo.[11] By 15 May 1945, a directive was issued to the US Army Air Forces requesting that Hiroshima, Kyoto and Niigata be put on a list of cities not to be bombed, so that they could be preserved as targets for a later date.[12] Nagasaki was not a priority site. The original target was an arms factory in Kokura, Japan, however weather conditions and residual smoke from a previous fire-bombing meant that the bomb was deployed to Nagasaki instead. Three bombing runs on Kokura were attempted, but 'at no time was the aiming point seen', as the flight log recorded.[13] *Bock's Car*, the bomber designated to drop the bomb on 9 August, gave up on Kokura and went on to its secondary target, Nagasaki.

A study commissioned by the *Asahi Shimbun* newspaper suggested that more than 57,000 people in Kokura – of a population then of 130,000 – would have been killed, and a central area two and a half miles in diameter would have been obliterated by the blast and by fire.[14] Kokura was the city that was meant to be bombed, that somehow survived. It was also the back-up target for the Hiroshima bombing, so if Hiroshima had been clouded over, the first atomic bomb would have been dropped on Kokura. The local community feel tied to the fate of Nagasaki victims. After the war, when they learnt that the bomb almost dropped on them, they realised that they could have been dead, or that their children might not have been born. It is significant that Kyoto was spared because of a personal intervention: the US Secretary of War, Henry L. Stimson, had fond memories of the place and did not want it bombed, demonstrating the arbitrary and personal lens between life and death in nuclear warfare.[15]

In the 1990s, Paul Lashmar interviewed pilots about dropping nuclear weapons on cities. He wrote that:

They dealt with this remarkably pragmatically, as military people do. They viewed it as a patriotic duty or as a job of work, retrospectively providing a successful deterrent. My questions stirred little reflexivity or rumination. I recall one pilot I interviewed, Colonel Sam Myers revealing for the first time his target: 'OK, my target for my crew was Gorky. And, this involved airborne alert missions. And we did have full weaponry aboard.'[16]

Historians and political scientists agree that the use of the atomic bomb was not necessary to gain surrender from Japan.[17] Robert Pape suggests that Japan was prepared to surrender at the time of the bombing; he writes, 'Japan's military position was so poor that its leaders would likely have surrendered before invasion, and at roughly the same time in August 1945, even if the United States had not employed strategic bombing or the atomic bomb.'[18] Therefore the bombings were executed in haste to ensure that they were still politically justifiable.

The 1946 War Department study *Use of the atomic bomb on Japan* supports the notion that Japan was ready to surrender anyway, mentioning that 'the Japanese leaders had decided to surrender and were merely looking for sufficient pretext to convince the die-hard Army Group that Japan had lost the war and must capitulate to the Allies'.[19] Hiroshima was bombed just two days before Russia declared war on Japan. This suggests that the bombing occurred in anticipation of Russia's declaration of war and, therefore, of Japanese surrender.

This implies that the USA had ulterior geopolitical motives for the bombing that were not predicated around saving lives, but rather demonstrating dominance over the Soviet Union.[20] The USA wanted to demonstrate multiple nuclear capacity – the ability to manufacture more than one nuclear weapon, and two different types of bomb were deployed to justify manufacturing both plutonium and uranium nuclear weapons. The scientists of Los Alamos anticipated a further series of deployments, so perhaps the end of the Second World War was the only thing that prevented the USA deploying further nuclear weapons.

The devastation of Hiroshima and Nagasaki was enabled not just by nuclear scientists and engineers, but by a prevailing culture of American anti-Japanese racism. American and British wartime propaganda animalised and dehumanised the Japanese people. Similarly to German Nazi propaganda, American anti-Japanese propaganda depicted a racist and anti-Japanese aesthetic, showing the Japanese as insects and vermin, to the point where 'an entire race of people appeared expendable'.[21] This dehumanisation allowed for the 'ethical' bombings of Hiroshima and Nagasaki, unnecessary to win the war, but creating a useful diplomatic advantage over Russia and eradicating a comparatively defenceless threat.

AMERICAN JUSTIFICATIONS

After the first atomic bombing, Allied forces dropped leaflets that said:

> The U.S. invented a powerful new atomic bomb. If you have any doubt about it, you should investigate the effects of a single atomic bomb on Hiroshima. One atomic bomb has the power equivalent to all bombs carried by 2,000 B-29 bombers. You should ask your emperor to stop this useless war by accepting 13 terms of honorary capitulation. Unless Japan gives up military resistance immediately, the U.S. are determined to use these bombs and other superior weapons to terminate the war quickly and powerfully.[22]

That leaflets were dropped upon the ailing population of Hiroshima, instead of medicine or humanitarian support, was a callous act that demonstrated the extent of mindless dehumanisation of the Japanese by the military leaders of the USA.

At home, however, the US attempted to minimise the public knowledge of the science and harmful consequences of Hiroshima and Nagasaki.

Information was distorted by campaigns of misinformation and reassurance; wartime censorship was maintained, and scientific experts such as Oppenheimer were brought out into the public domain to incorrectly reassure people. Public and academic knowledge of the effects of ionising radiation became caught up in a process of 'agnotology', the cultural production of ignorance.[23] Notably, 132 articles relating to the Hiroshima and Nagasaki bombings that were produced soon after by *The New York Times* omitted any information about ionising radiation and its effects on bombing victims.[24] Despite this effort, it was impossible to suppress all knowledge of the consequences of the bomb. Searing journalistic accounts by John Hersey in the *New Yorker* and by Wilfred Burchett in the *London Daily Express* described the unusual and painful conditions of death that occurred for the victims of Hiroshima.[25,26] When European newspapers raised concerns about radioactivity, this was suppressed as propaganda. It didn't affect the American public's perception of the bombings. People in the US were already war-worn, and a study conducted after showed that 85 per cent of those surveyed were in favour of the atomic bomb.[27]

The science of the atomic bombings remained cloaked in secrecy after the Second World War. The publicity surrounding the consequences of the bombings – blast, fire and radiation – were each treated quite differently.[28] While the power of the blast was proudly described within the euphemism of powerful conventional bombing by President Truman, other harmful aspects of the atomic bomb received considerably less attention. Fire damage was not investigated. The public already had knowledge that bombing caused fires, following the firestorms created by Allied bombing of Dresden and Hamburg, and the firebombing of Tokyo. It took time and information for experts to apply their understanding of conventional bombing to the new atomic scenario.

The aspect that was actively suppressed to the American public was the consequences of high-level ionising radiation exposure. Military, industrial and scientific leaders were dragged into the limelight to try and ease fears.[29] In the long term, ionising radiation has been the most noteworthy aspect of the atomic bombs, yet this was the least publicised and least understood aspect of the bomb's effects. Circulation of medical reports was limited, and newspapers were censored in an attempt to quieten outspoken individuals.

Japanese scientists collected evidence and studied the symptoms of those who were ill and dying of acute radiation syndrome (ARS).[30]

However, they were not provided with information about the radioactivity by the US or UK. Initially, a chemical or biological agent was suspected as the cause of this illness and fatalities. The idea of a 'germ bomb' was initially posited by doctors. It took weeks for scientists to discover that death was due to immune system damage, when decreased white blood cells and platelets were noted in late August. However, there was no universal terminology for the effects of ionising radiation from atomic weapons, so reports of this time refer to symptoms as 'atomic bomb illness', 'atomic poisoning' and 'atomic plague'.[31] To create further complications, Japanese reports and case notes were confiscated and sent to the USA, where they remained classified for years at the Armed Forces Institute of Pathology. Thus, medical military personnel with security clearance in the USA were able to develop an in-depth understanding of the biological effects of the bombs without providing the greater scientific or medical community, let alone the public, with any insights. The motivations for this censoring of information included disguising US scientists' own lack of knowledge of the consequences of the bomb. Additionally, government officials who made decisions about the bomb wanted it to be viewed as a regular and just weapon, used to end a just war. No one in the USA wanted the bomb to be linked to chemical or biological warfare, which was stigmatised within both public and military domains.[32]

Not everyone subscribed to the propaganda campaign. The work on the atomic bomb joined science and national security together, but scientists were left compromised. After the first atomic bombings, Oppenheimer expressed qualms about the next stage of nuclear defence, the hydrogen bomb. He became a pariah within the Los Alamos community, losing any power and respect as an atomic scientist.[33] Oppenheimer was positioned not only to shape, but to also to be shaped by the insular Los Alamos community. Oppenheimer had his security clearance revoked by 1946, but he remained the director of Princeton's Institute for Advanced Study from 1947 until the end of his career, despite these perceived transgressions.[34]

THE DEMON CORE

The devastating after-effects of the bombing of Hiroshima and Nagasaki were visible. However, the effects of nuclear weapon testing and production initially appeared to be positive for workers in the scientific community of the Manhattan Project, outwardly providing stable careers and opportunities for intellectual creativity for them. The dark, secret history of the place only emerged later, as the legacy of harm to nuclear weapons workers was gradually revealed.

There is considerable profit for the designers and manufacturers of nuclear weapons, because of the levels of expertise and skill required, but also due to the risk and stigma associated with working with nuclear weapons. Around half of the people employed on the Manhattan Project were engaged in manual work, and many injuries occurred among employees due to standard occupational risks, rather than ionising radiation exposure. There was too much for any one person to understand, and risks were nascent, waiting to emerge. Some risks were banal, including injuries from heavy machinery and injuries on construction sites. Twenty-four fatal accidents occurred at Los Alamos alone during the design and manufacturing of the first bombs, from 1943 to 1946. These included people crushed by rocks, piles of steel and crankshafts, as well as accidental shootings, vehicle crashes, drownings, fatal horse-riding injuries, barbiturate overdoses, alcohol poisoning, and exploding smoke bombs.[35] An example of the more exceptional historic challenges that have arisen in Los Alamos health and safety culture is the cautionary tale of the *demon core*. It is the tale of a nuclear physicist who was literally destroyed by his own creation.

The demon core was a small but subcritical mass of plutonium involved in two accidents. In the context of nuclear weapons, this is where the core of an artificial actinide, such as plutonium, is bombarded by neutrons, allowing radioactive energy to be rapidly released. This reaction is dangerous by design, but can be controlled or uncontrolled, where an uncontrolled criticality can result in exposure to ionising radiation and harm. The process was known by the Los Alamos scientists as tickling the dragon's tail, due to its inherent danger.[36]

The demon core was originally intended for a third bombardment during the Second World War, so was designed with a tiny margin for safety to ensure a successful detonation. Perhaps the demon core's moniker is a misnomer, as the core should have been controlled and managed appropriately. Instead, the harm that arose was due to poor health and safety practices in Los Alamos during the post-war era of very rapid nuclear weapons development and testing. The harmful effect of exposure to ionising radiation had been known since female radium dial painters in the 1920s became unwell with anaemia. This resulted in health challenges, including bone fractures and necrotic radium jaw from licking radioactive paintbrushes to paint the dials precisely.[37] However, the nuclear scientists of Los Alamos considered themselves to be cavalier and daring pioneers of their new trade, and were willing to take enormous risks to produce new discoveries.

The two incidents occurred on 21 August 1945 and 21 May 1946, resulting in the deaths of two Los Alamos employees.[38] During the first incident, a scientist called Harry Daghlian was working alone, performing neutron-reflector experiments on the core. The core was held within a stack of neutron-reflective tungsten carbide bricks, and the addition of each brick brought the assemblage closer to criticality. Daghlian accidentally dropped a final brick directly onto the core, creating a self-sustaining critical chain reaction known as a supercriticality, and exposing himself to a high dose of ionising radiation. He died a few weeks later from ARS, on 15 September 1945.[39]

The second incident occurred when Louis Slotin was demonstrating to his colleagues how neutron reflectors could be used to identify the precise point at which subcritical plutonium becomes critical. The experiment required that two half spheres of neutron-reflector beryllium were placed around the demon core, while a top reflector was lowered above the core, through a small hole. However, allowing the reflectors to close completely would result in the immediate formation of critical mass. The only thing that prevented this in Slotin's experiment was the blade of a standard flat-headed screwdriver, held in his other hand. On the day of the accident, Slotin's screwdriver slipped outward a little, allowing the top reflector to fall into place around the core. The core became supercritical and released an intense burst of neutron radiation, a deathly blue flash. Slotin quickly flipped the top shell to the floor. This stopped the criticality, but Slotin received a lethal dose of radiation in the process and died nine days later, on 30 May 1946.[40]

LOS ALAMOS LADIES

The story of Los Alamos is often described in a way that focuses upon the successes of the male scientists and military officials who worked on the bomb, rather than the experiences of women. However, as with other scientific breakthroughs of this era, such as codebreaking at Bletchley Park, there were many women who worked at Los Alamos during the race for the atomic bomb. Among this hidden human geography of women were scientists, engineers and mathematicians who worked at Los Alamos Nuclear Laboratory, as well as housewives, cleaners and teachers who supported the community.[41] Several hundred women worked on the Manhattan Project. These women worked hard to support the development of nuclear weapons, though they received little credit at the time.

While the Manhattan Project is no longer surrounded by secrecy, there is still little information about the contributions of women within the public domain. We know that no women rose through the ranks to become project leaders or commanders.[42] In fact, one of the motivations for employing women scientists was due to the high risk of failure that surrounded work on the atomic bomb. Many successful male scientists and engineers were assigned to projects that were perceived as more likely to succeed.

Some women have been recognised for their work. These women include physicist Leona Woods Marshall, a member of Fermi's research group and an expert in detecting and monitoring neutron flux.[43] She was an important participant in the first self-sustaining nuclear chain reaction experiment, proving that nuclear energy had the potential for explosive release. Other female scientists included the experimental chemist Isabella Karle, who worked on the chemistry of transuranic elements; and Lilli Hornig, a chemist at Los Alamos.[44] When Hornig was asked to take a typing proficiency test, she said 'I don't type' and instead explained her formidable credentials.[45] She became a staff scientist and worked with plutonium salt chemistry before leading the development of explosive lenses – the 'outer packaging' of a nuclear weapon, used to compress the core fissionable material of a nuclear weapon and induce a critical state.[46] After the Trinity test, she petitioned to prevent the bomb being dropped in warfare, but her campaign was ignored.

Perhaps the most significant female scientist of Los Alamos was the physicist Maria Goeppert Mayer, who had fled from Nazi Germany in the 1930s.[47] Mayer initially worked on nuclear shell structure, then began to explore the thermodynamic properties of uranium hexafluoride gas, pioneering the use of this gas to enrich uranium by separating U-235 isotopes from U-238 isotopes. After the Second World War, she worked on the development of the hydrogen bomb. In 1963, she became the second woman, after Marie Curie, to win the Nobel Prize in physics for her work.

These women lived and worked in Los Alamos alongside male scientists and their wives. Many of the women of Los Alamos moved there early on in their marriages, with their husbands. Topics that they discussed with their husbands during their doctoral years became taboo subjects. Many of the women found this secrecy very difficult, and disruptive of their relationships with their husbands. Los Alamos wives had their mail censored and couldn't even subscribe to magazines for reasons of security. There were complaints from the USA army about the cost of maternity bills, as young families blossomed.

After the war, many of these brilliant women were no longer needed, and were discouraged from continuing their careers as scientists and academics when men returned home and needed jobs. Their contributions were sidelined and not memorialised in the same way as the men of Los Alamos. In fact, the only woman who gained publicity for her work was Lisa Meitner, a physicist who was never actually involved in the Manhattan Project. Many of these women who had provided a significant contribution to nuclear science left their work at Los Alamos to become housewives and mothers. However, some women were tenacious enough to ignore the propaganda and pursue their own careers as academics across the USA. It wasn't just women who lost their jobs or retired – over 10,000 people in total had left the Manhattan Project by September 1945.

A LONG RECOVERY

When the bombs were deployed on Nagasaki and Hiroshima, there were immediate and massively harmful consequences. Thousands of people died from the intense heat of the blast. Many others suffered burns and radiation injuries. Makeshift hospitals were overcrowded as they sought treatment for their wounds. The bombs created an instantaneous wave of heat 3,000–4,000°C at ground level. Immediately after the bombing of Hiroshima, there was 'nothing but silence and desolation'.[48] Buildings were razed to the ground, windowpanes shattered, the urban geography of Hiroshima was irrevocably changed. The blast had destroyed more than 10 square km of the city, and the intense heat caused fires that spread across the remains for over three days. 'About 90% of the city's 76,000 wooden buildings were partially or totally incinerated and concrete was reduced to rubble. Of the 33 million square metres of land considered usable before the attack, 40% was reduced to ashes.'[49] The city was no longer Hiroshima, having become an ashen blast zone within minutes, the entire city destroyed by a bomb. Even underground air raid shelters with earth-covered roofs were destroyed.

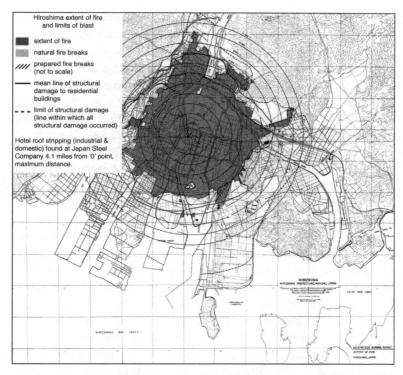

Figure 2.1 US military damage maps produced in 1946 demonstrate the extent of damage to Hiroshima and Nagasaki

Source: US Department of Defence

Despite being wiped off the map, Hiroshima was a city of determined survivors. A huge civic effort was undertaken, aided by the arrival of large numbers of volunteers from across Japan. Yuki Tanaka, a retired historian of Hiroshima City University said that 'Hiroshima received a lot of help from people in neighbouring towns and cities such as Fuchu, Kure, and even Yamaguchi.'[50] This collaborative effort went a substantial way towards changing the initially bleak fate of Hiroshima.

It was a significant challenge to organise the cremation and burial of the dead, and this became an ongoing process during the aftermath of the bomb. However, businesses and organisations were determined to restore basic amenities to the residents of Hiroshima. The Bank of Japan reopened two days after the bombing, offering floor space to other banks with destroyed premises, and with tellers working under open skies in good weather and under umbrellas when it rained. A limited rail and streetcar service resumed on 9 August, the same day that Nagasaki was bombed. Water pumps were repaired and started working again four days after the bombing, although damaged pipes leaked water, creating mud out of the ashes of destroyed wooden homes. By 14 August, experimental phone lines were back in operation, despite the bomb annihilating every employee of the central telephone exchange. Power was restored to all houses that had escaped fire damage by November 1945. Higashi police station was used as a base for search and rescue operations. However, as steps towards recovery began to be made, a powerful typhoon deluged the city on 17 September 1945, flooding temporary hospitals and relief centres. The typhoon delayed the reconstruction of the city, but also swept much of the radioactive fallout from the bomb into the sea, where it dispersed across the ocean.[51]

Survivors of the bombings were labelled *Hibakusha*, meaning 'the exposed'.[52] Inequalities were created, due to a shortage of places for people to live. Demand for housing turned the region surrounding the epicentre into a shanty town of wooden shacks in the aftermath of the bomb. There were already hidden social inequalities across Hiroshima before the bombing. Communities of migrants, including Korean labourers and 'comfort women', lived in wooden shacks and were members of the *burakumin*, or underclass.[53] These inequalities were intensified after the city was bombed, and within months thousands of people were living on the riverbanks with no access to sanitation or electricity. These huts remained until new high-rise flats were built in the 1970s.[54]

The stigma surrounding radiation exposure created new challenges. Fears and misunderstanding of the health and genetic effects of ionising radiation meant that people who had survived the bombings became pariahs beyond their communities, facing bleak and friendless futures. If you had been involved in the bombings, it was difficult to find a partner or get married. This stigma continued for several years. Sunao Tsuboi described the social difficulties of being *Hibakusha*, explaining that his fiancée's parents refused to let her marry him, due to concerns about how long he would live.[55] They were only able to marry once his future parents-in-law realised that he wasn't going to die young.

Reconstruction replaced warfare, and local authorities launched a five-year recovery plan. A plan for a memorial to the bombing gained traction when the local *Chugoku Shimbun* newspaper launched a competition in 1946, seeking readers' visions for the city. The peace activist and atomic bomb survivor Sankichi Tōge won this competition, after suggesting an international peace memorial garden with a museum and library.⁵⁶ However, the city was financially ruined after the bomb, as tax revenues had plummeted. In 1949, national recognition of the need for redevelopment led to the passing of the Peace Memorial City Reconstruction Law, providing new sources of funding for reconstruction, and freely donating land that previously belonged to the military.⁵⁷

Figure 2.2 Labourers working on the restoration of Hiroshima's Aioi Bridge in 1949
Source: Yoshita Kishimoto

The reconstruction of Hiroshima began without constraints of land ownership and funding, supported by the blossoming Japanese post-war economy. City planners faced a dilemma: how much of the ravaged city to preserve for posterity. The *Hibakusha* didn't need constant reminders of the tragedy that they had experienced, so it was decided to preserve just one building as a monument within the park. This was the A-bomb dome, previously the Industrial Promotion Hall. Its tattered girders splayed like ribs, chunks of stone surrounding it, it is a brutal and visceral reminder of the harm that came to Hiroshima. The ruined dome was surrounded by a graceful 30-acre parkland site not far from Ground Zero, designed by the famous late Japanese architect Kenzo Tange and completed in 1954.⁵⁸ The A-bomb dome became a UNESCO world heritage site in 1996, more than fifty years after the bombing.

Meanwhile in the USA, Project Ichiban was under way by 1962. This Los Alamos study created a first mock-up of Little Boy to try to recreate and understand the blast effects of the Hiroshima bombing in a controlled environment. In 1982, a further re-creation of the original bomb was used for neutron emission experiments and to reassess the yield of Little Boy.[59] No lessons had been learnt.

A TRANS-PACIFIC RECOVERY

Los Alamos is still a place of contradictions and secrets. It is a *plutopian* idyll, and the birthplace of the atomic bomb. It is a pristine and beautiful town of white picket fences, surrounded by the dusty canyons of the New Mexican mesa. This federally funded community includes some of the brightest physicists and mathematicians in the USA among its residents, well-paid and highly educated, while the impoverished local Hispanic residents of the surrounding pueblos sell cheap knickknacks to tourists. The inequality between Los Alamos and surrounding communities demonstrates that trickle-down economics is not working, as it seems that neighbouring communities have not benefited from the presence of the town. Los Alamos National Laboratory is still a breezy campus of nuclear secrets, separate from the world around it, purporting to be 'Delivering science and technology to protect our nation and promote world stability'.

Figure 2.3 Los Alamos, a place of discoveries and secrets
Source: Photo by B. Alexis-Martin, 2016

Los Alamos will continue to be a centre of social and environmental justice issues, as much as a centre for nuclear defence for the foreseeable future. It is still a place of secrets and paradoxes. There is still complicated relationship between the academic and local communities of Los Alamos. Despite research transitioning to include climate change, supercomputing, materials science and astrophysics, nuclear defence research remains at the forefront of the current agenda. George Orwell may well have been proven right when he described the inequalities created by the American bomb in 1945, as:

> robbing the exploited classes and peoples of all power to revolt, and at the same time putting the possessors of the bomb on a basis of military equality. Unable to conquer one another, they are likely to continue ruling the world between them, and it is difficult to see how the balance can be upset except by slow and unpredictable demographic changes.

Recently, these changes have begun to emerge and a realistic pushback against nuclear weapons is under way.

The demographic positions of the US and Japan have reversed, with Japan now having greater equality and some of the longest life expectancies in the world, whereas the US has one of the lowest life expectancies and the greatest inequalities for an economically highly developed nation. Visit Hiroshima now, and the geography and demography has changed. Japan has undergone reconstruction and redevelopment, booms and busts, and Hiroshima has remained a city of beautiful parks and tall skyscrapers. It has a population surpassing 1.19 million, a burgeoning gourmet scene, towering luxury shopping centres and a trendy night life. It is a city of vibrant green boulevards and open spaces, entangled by the braided tributaries of the Ōta River. However, it is also a city of memorialisation. While the city grows and evolves, the memory remains of Hiroshima as the first place on Earth where nuclear weapons were used in warfare, on 6 August 1945. The A-bomb dome and other memorials, large and small, scattered across the city as if by the wind, are the most significant reminders of Hiroshima's legacy. Adult survivors of the bombing are now aged pensioners, and as their numbers decline, so do the number of keepers of personal testimony and memorial.

Japan is a nation of stories, and the atomic legacies of weaponry and accident are interwoven into its cultural fabric through the tales of the

Kataribe. This is a form of spoken word storytelling that previously focused upon legends of mythical creatures, monsters and dragons to describe history, society and culture. While the purpose of Kataribe remains the same, the structure and design of the storytelling process has adapted since Hiroshima and Nagasaki to include tales of the bombings. The *Hibakusha* community maintain a living collective memory of the bomb, sharing their atomic folktales as a cautionary modern mythology against nuclear war.

Unfortunately, not all *Hibakusha* have been recognised. A select number of survivors who experienced both bombings have had specific difficulties with gaining this recognition. Tsutomu Yamaguchi was unique in having finally being recognised as surviving both the Hiroshima and Nagasaki bombings. A Nagasaki resident, he was on business in Hiroshima on 6 August when the city was bombed, and he was injured. Despite his wounds, he travelled home and reported to work in Nagasaki on 9 August. He was in the process of describing his experiences of the previous blast to his incredulous boss, at the moment when Fat Man was detonated and Nagasaki was destroyed. He finally received recognition of his experiences by the Japanese government on 24 March 2009. He passed away from stomach cancer at the age of 93 on 4 January 2010, and upon his death the Mayor of Nagasaki said that 'a precious storyteller has been lost'.[60] Tsutomu Yamaguchi was known for saying 'My double radiation exposure is now an official government record. It can tell the younger generation the horrifying history of the atomic bombings even after I die.' His emotions mostly emerged in *Tanka*, or 31-syllable poems. He wrote hundreds, each one an ordeal. When he composed them, he would dream of the dead lying on the ground. One by one, they would get up and walk past him. Here is an example of one of his Tanka poems.

> Carbonised bodies face-down in the nuclear wasteland
> all the Buddhas died,
> and never heard what killed them.
> Thinking of myself as a phoenix,
> I cling on until now.
> But how painful they have been,
> those twenty-four years past.

The 183,500 registered *Hibakusha* of Hiroshima and Nagasaki receive a monthly allowance and medical care from the Japanese Red Cross

Nagasaki Genbaku Hospital, which was established after the bombings. Many are succumbing to illnesses that are associated with old age, but could also be connected to exposure to ionising radiation. This makes it difficult to determine the true long-term health impacts of the bombings.

HIROSHIMA MON AMOUR[61]

It was assumed that nothing would grow within the bleak 1.6 square km blast zone for 75 years. However, fresh stems quickly pushed through the damaged earth, plants took root, and the branches of the *Hibaku-jumoku*, the survivor trees, unfurled leaves of weeping willow and oleander from budded stalks. The city has been rehabilitated and it is challenging to imagine it as a place of devastation. Hiroshima Peace Memorial Park is a lush focal point of this re-greening process, and a unique human ecosystem has sprung up among the gingko trees and susurrating cicadas.

The park has its own distinctive psychogeography, providing a public space for complex emotions and experiences to be explored by locals and tourists. International visitors feature prominently around the larger memorials and cenotaph. They ring the delicate origami crane bell-pull within the Children's Peace Monument, take a few photographs of the cenotaph, stroll beside the Peace Pond, and then across the river to the A-bomb dome. Distance is no indication of personal connection, and victims of Hiroshima originated from across the USA, China and South East Asia. Thousands of Koreans died in Hiroshima. The men were forcibly conscripted and the women performed the duties of 'comfort women'. The monument and Cenotaph to Korean Victims are both festooned with brightly coloured flowers and receive a constant trickle of visitors, many of whom are Korean. Swags of peace cranes garland the smaller memorials dotted about the park, and the fragrance of sandalwood and citron lingers, as incense is lit and local heads are respectfully bowed. Japanese schoolchildren come here to learn, and they sit in the shade of the trees at noon in civilised huddles, to eat lunch and chatter. However, there are places and spaces that cannot be entered, and the park itself is carefully managed and controlled.

Many visit to reflect upon the atrocity of the bombing, but this attitude is not universal. I learnt this during an encounter with an American man at the Ground Zero memorial, tucked away on a side-street beyond the boundaries of the park. We smiled at each other, as he shared his reasons

Figure 2.4 Public incentives not to investigate the A-bomb dome too closely
Source: Photo by B. Alexis-Martin 2017

for visiting, declared the power of the bomb to end the war, and the American soldiers, including his grandfather, whose lives were saved by this action. He was grateful for the bomb, but I was shocked at the way he had decided to make an emotional connection with this place. This also exemplifies how, for some, Hiroshima's adoption of peaceful aims is seen as an example of denial of its aggressor status during the Second World War.

The local community has a deep and profound connection to the park. Volunteers in distinctive uniforms meticulously maintain the place daily. This voluntary care of space intensifies as Hiroshima Peace Day draws near. Visit the park at 6 a.m. towards the end of July and you will discover hordes of elderly people from the 'Senior University', wearing sunhats and brandishing trowels. They crouch above the ground, plucking weeds from the soil with gloved fingers. While they garden, trails of elegantly dressed office workers bisect the park at intervals, carrying files and parasols in delicately gloved hands. Commuting to work, this stretch of land has become another familiar part of the rhythm of their daily lives.

Strangely, there are also spaces of conflict and deviance here. The *Uyoku dantai* are the Japanese extreme far right. They call themselves the Society of Patriots and travel about in dark vans painted with worrying slogans. War crime denialists, they support historical revisionism, oppose socialism and want Japan to join the nuclear circus. They cannot be arrested; freedom of ideology is protected by the Constitution of Japan. So, they jeer from the sidelines of the park, and organise protests outside the A-bomb dome on Hiroshima Peace Day. Like other far-right groups globally, they have been gaining popularity in recent years.

Figure 2.5 Some of the Hiroshima Society of Patriots protesting outside the Peace Park on Peace Day, 6 August. They did not like me as much as this photo suggests

Source: Photo by B. Alexis-Martin 2017

Conversely, there is also a place of joy hidden within this park, on a dusty corner of dry earth behind the public toilets. Here, a group of elderly Japanese men meet every week-day morning, to crouch on battered wooden chairs and play board games. Some, but not all, are *Hibakusha*, but all of them look relaxed, and laugh loudly as they engage in drawn-out battles of Shogi and Go. They have created their own friendly yet private space within this park. As dusk sets in, they pack up their board games, fold up their little chairs and tables and go home. The cicadas grow louder and a calmness settles over the park as twilight descends. Small clusters of local teenagers gather and relax in the evening's warmth. Faint sounds of conversation gradually dwindle to nothingness and the day ends, reclaimed by the stillness of night.

3
The Mystery of the X-ray Hands

> Well that was certainly a lot of bang from just a little plut ...
> – Sir William Penney

The devastating aftermath of the atomic bombings of Hiroshima and Nagasaki is undeniable. Rather than consigning these events to history, other global superpowers wanted to own a piece of this atomic muscle. The demise of empire and the considerable debt burdens of states, including France and the UK, after the Second World War meant that nuclear weapons appeared to be an enticing fix for them to regain prominence on the world stage after relinquishing many of their colonies. A new era of rapid nuclear proliferation began, and nations began to test and stockpile weapons in earnest. Atomic bombs evolved into hydrogen bombs and thermonuclear warheads. These weapons were tested in the atmosphere and underground. The effects, particularly of atmospheric testing, have been global in nature.

Here, I consider the experiences, and health and safety implications of nuclear weapons testing for the soldiers and scientists who undertook this work, and their families. These people are known as the *atomic veterans* and *nuclear test veterans*.[1] Some have become radical pacifist activists, others stay loyal to their governments despite enduring health and social challenges. Here, colonialism, human rights abuses, cultures of secrecy and disconcerting approaches towards the health and safety of nuclear test participants are scrutinised internationally. The spaces and places of activity, activism and memorialisation of the atomic veterans are also explored.

The experiences of the atomic veterans were shaped by the conditions, risks and consequences of life in the nuclear military industrial complex. A haphazard health and safety culture combined with a culture of secrecy has had a long-lasting impact upon these men and their families.[2] Although the veterans believe that it was the tests that posed the greatest risk, their work presented many other significant hazards. The climate and geography of the isolated places where tests were conducted had a profound effect upon the mental and physical health of servicemen, and basic living conditions worsened these effects. Health challenges beyond ionising radiation included industrial accidents, the extensive use of carcinogenic DDT on the troops, poor sanitation, dysentery, severe sunburn and inadequate rations.[3] The way that these risks were faced and managed has shaped the veterans' understanding of their time working on nuclear test series. Despite governments' assertions that the nuclear tests involved little or no risk of radiation exposure, they have had significant repercussions for atomic veterans and their families. The true health, psychosocial and cultural costs of the tests are only beginning to emerge. The chapter concludes with a table of statistics (Table 3.1) that reflect the outcomes for these veterans internationally.

Nuclear weapons testing was undertaken in a few remaining colonies. They were isolated, far from home, out of sight and out of mind, except when successful tests were reported to the media with pride and bombast. Island and desert outposts were used, far away from 'civilised' humanity, and barely registering on the map to most people. The colonial geographies of the nuclear weapons tests were reported as uninhabitable wilderness by the senior military officers who chose them. This was often far from the truth. Local communities were forced from their homes and sacred lands at best; or they were left there to become human guinea pigs at worst, potentially exposed to high doses of ionising radiation.

The men who tested the nuclear weapons were from the countries that later became the five nuclear weapon possessor states: the USSR, UK, US, France and China. These men were often young and uneducated, with little information about the risks they could face.[4] Many of them were undertaking their national service, or a similar conscription programme. In an era when travel was very expensive for the average person, testing nuclear weapons offered them unimaginable opportunities to see the world. They travelled away from the social regulation of tightly knit family and life-long friends, away from everything familiar, while jumbled up into regiments with a random assortment of other soldiers. This was the first time that many of these men would be able to define themselves on their own terms. There are issues of medical, social, cultural and environmental justice surrounding their experiences, as for those other men who were involved in the *remediation* of these atomic places, attempting to restore the original geography and ecology, and trying to remove traces of this nuclear attack on the land.[5]

JUST TESTING

Nuclear testing is often portrayed as a bombastic process, overshadowed by the devastating majesty of mushroom clouds and described in terms of annihilation or dramatic uncertainty. This is understandable, given the inherently destructive nature of nuclear weapons. The stories of the daily routines and experiences of the men who work on the tests are often therefore neglected, in favour of portraying the more dramatic might of the bomb.

Some of these international tests were conducted by the US government on the Bikini and Eniwetok Atolls in the Marshall Islands, Kiritimati (formerly Christmas) Island, Johnston Island and Kalama Atoll.[6] The US Atomic Veterans were those who participated in atmospheric and underwater nuclear weapons tests from 16 July 1945 to 30 October 1962. In total, 1,066 atomic weapon device detonations were supported by the US Department of Defence and the US Atomic Energy commission between 1945 and 1992.[7] There were consequences of these detonations to indigenous, civilian and military communities alike. In the USA alone, approximately 500,000 military and civilian personnel were involved in testing.[8] This included personnel at all levels from conscripted military workers to the scientists of defence laboratories like Los Alamos.

The purpose of these tests was to develop and enhance the US nuclear weapons arsenal. Despite attempts to hide the harmful effects of the tests, evidence of accidents and exposures has emerged. These included the exposure to high levels of radiation of Marshallese and Japanese fishermen following the US Castle Bravo test in the Marshall Islands on 1 March 1954.[9] It was the highest yield American test, exploding with a force equivalent to 15 million tonnes of TNT; and was 25 times more powerful than expected.[10] Proving negligence or harm from nuclear weapons tests is usually difficult. Information about the human impact of the tests has been withheld, obscured by technical language, or never collected in the first place. However, the Castle Bravo test remains the most significant radiological incident in American history.

The UK, China, France and the Soviet Union also conducted nuclear weapons tests.[11] The first Soviet atomic bomb was tested in 1949.[12] The Soviet Union tested its nuclear weapons in regions far from Moscow, most notably the Semipalatinsk test site in Kazakhstan, and Novaya Zemlya, a northern archipelago of Russia. Five hundred indigenous people were removed from Novaya Zemlya to make way for the test site for 'Tsar Bomba' – the largest hydrogen bomb ever tested.[13] The Tsar Bomba was 3.3 times more explosive than Castle Bravo, America's largest H-bomb test. Further, the 200,000 residents of the Semipalatinsk *oblast* or region were used as 'human guinea pigs' to study the effects of radiation in Soviet Kazakhstan. Tests were undertaken in Semipalatinsk for over forty years, resulting in severe environmental and long-term health effects.[14] It took mass protests in Kazakhstan to drive the Kazakh government's decision to close down the Semipalatinsk nuclear testing site in 1989.[15] The UK tested in Aboriginal homelands of Maralinga, Emu Field and the Montebello Islands, in addition to Malden Island and Kiritimati.[16] British testing had a disproportionate impact upon indigenous people, many of whom continued to move throughout the region at the time of the tests. It was later discovered that a traditional Aboriginal route crossed through the Maralinga testing range. The effects were not only radiological; restrictions on the indigenous population's access to their traditional lands also caused psychosocial and cultural problems.[17]

The US and UK did not initially cooperate with each other during the first phase of post-war nuclear defence development. Instead, each country's tests were undertaken as solo endeavours, with the UK successfully developing and testing its first hydrogen bombs (H-bombs) during Operation Grapple on 8 November 1957. Meanwhile, the USA covertly supported the French nuclear defence programme. This defensive stalemate between allies ended on 3 July 1958 when the US–UK Mutual Defence Agreement (US–UK MDA) was signed. The special relationship that the US–UK MDA initiated was apparently beneficial to the UK in terms of expertise and defence capacity, but it was never a relationship of equals. The USA had much greater military and economic resources than the UK and was able to test its nuclear weapons in remote Nevada deserts. In some ways this was an ideal location, with stable geology and comparative isolation; this meant that future joint tests could be moved to the USA. The UK soon became dependent upon the USA's nuclear deterrence expertise.

This is a situation that persists to this day, with the most recent renewal of the mutual aid contract extending until 31 December 2024. Decisions that are made for the UK's Trident nuclear deterrence system are directly influenced by the USA, preceded by the historic decisions of presidents from Truman to Reagan and, most recently, Trump.

Table 3.1 The UK nuclear tests undertaken during the atmospheric testing series, including numbers of participants for each operation

Service/employer	Rank	Test participants				Controls			
		National servicemen	Regular	Total no.	%	National servicemen	Regular	Total no.	%
RN[a]	Officer	54	434	488		22	559	581	
	Other ranks	340	5,477	5,817		261	6,502	6,763	
	Total	394	5,911	6,305	29.5	283	7,061	7,344	32.9
Army	Officer	24	537	561		174	488	662	
	Other ranks	1,563	3,670	5,233		1,727	3,093	4,820	
	Total	1,587	4,207	5,794	27.1	1,901	3,581	5,482	24.5
RAF	Officer	17	1,594	1,611		43	1,755	1,798	
	Other ranks	404	6,428	6,832		765	6,139	6,904	
	Total	421	8,022	8,443	39.5	808	7,894	8,702	39.0
AWE[b]	Social class 1	0	380	380		0	361	361	
	Other social classes	0	435	435		0	444	444	
	Total	0	815	815	3.8	0	805	805	3.6
All services and employers	Total officers/social class 1	95	2,945	3,040	14.2	239	3,163	3,402	15.2
	Total other ranks/social classes	2,307	16,010	18,317	85.8	2,753	16,178	18,931	84.8
	Total	2,402	18,955	21,357	100.0	2,992	19,341	22,333	100.0

Notes: [a] RN includes members of the RM, RNVR and NAAFI; [b] AWE includes a few employees of AERE Harwell
Source: Muirhead et al. See note 31

BOMBS AWAY!

In total, the British government conducted 64 nuclear weapons tests in Australia and the Pacific Islands between 1952 and 1963, in an attempt to prove that the country was as technologically advanced and as worthy of significance as the USA and the USSR.[18]

Seven of the nuclear weapon tests took place in Maralinga, a remote area of South Australia whose name originates from the Aboriginal Yolngu term for 'thunder'. Maralinga did not host major H-bomb tests, instead providing space for atomic bomb tests and the testing of the components for thermonuclear weapons.

Many of the British and Australian servicemen involved in the testing experienced perceived health and reproduction problems, because of exposure to radiation and other environmental risks. One veteran in the Nuclear Families study commented that he 'got radiation twice', adding: 'I had a good scrub and a tablet and [was] told to bugger off.' He described losing his first child, explaining that 'when it was born it was a lump of meat ... you couldn't tell if it was a boy, girl, human, whatever'.[19] The serviceman was given no compensation and, upon returning to work after five weeks' leave, was told by his chief engineer to 'just get on with it'. However, it is important to discern between the perceived immediate deterministic health effects and stochastic long-term health risks. While there have been health challenges for the veterans, studies have shown that there are no known long-term hereditary effects from short-term moderate-dose ionising radiation exposure.[20] However, the environment at Maralinga presented a plethora of immediate risks. A serviceman recounted that three of his colleagues died because of a tropical bug. 'They got big spots all over them, big sores and that was it, they passed away,' he said. This is a plausible scenario, as is the possibility of exposure to other diseases and environmental toxins, including beryllium and DDT.[21]

Another British Maralinga veteran describes his experiences as follows:

> It was 1959 when I arrived in Australia aged 19, for what seemed like a great adventure. I was stationed at RAAF [Royal Australian Air Force] Edinburgh Field just north of Adelaide. Part of the station's minor trials section was at Maralinga Atomic Range, my detachment to Maralinga came soon after I arrived. While stationed at Maralinga, we just went

about our normal duties with the addition of servicing vehicles and stationary units in the forward area. This did not mean much at the time, as all major tests had been transferred to Christmas Island, locally known as Kiritimati. Minor trials were carried out during my time there. It was only when the Royal Commission reported on the Atomic Tests that were carried out in Australia, that the full significance of the term Minor Trials came to light.

Since the conclusion of the nuclear tests at Maralinga in 1963, there have been several clean-up operations. The first, Operation Brumby, took place in 1967, and has been described by Professor John Keane as a 'quick brooms-around-the-toilet-floor effort by British army engineers'.[22] Keane also claims that the attempted clean-up 'scattered and left behind a great deal of radioactive material'. Indeed, in 1980, surveys conducted by the Australian Radiation Laboratory revealed that the site was not safe, prompting a further decontamination operation. This was completed in 2000, but there has since been much criticism from experts, including Alan Parkinson, a nuclear expert and former advisor on the clean-up operation, who stated in 2002 that poor management and execution of the project had 'left hundreds of square kilometres of Aboriginal lands contaminated and unfit for re-habitation'.[23] Despite this, the Australian government claims that the project was successful, and as such most of the test site is now safe – with the exception of a 120 square km area which is deemed uninhabitable for the foreseeable future.

BOMB GONE

Between May 1957 and September 1958, the British government tested nine thermonuclear weapons in the Pacific – Operations Grapple, Grapple X, Grapple Y and Grapple Z.[24] This was followed by the Operation Dominic series of 31 detonations, conducted on Christmas Island, Kiribati and Johnston Atoll by the USA with support from the UK in 1962, after a UK–USA test moratorium in preparation for the Mutual Defence Agreement from 1958 to 1961.[25]

This afflicted region became known as the Pacific Proving Grounds.[26] Christmas Island, now Kiritimati, was a coral atoll in the Pacific Ocean that served as the base for the servicemen, scientists and civilians involved in Operation Grapple and Dominic. The success of the Grapple tests cemented Britain as a thermonuclear power, but their negative impact upon the servicemen stationed on Christmas Island was similarly significant.

The British government has described test procedures and safety measures as 'meticulous', but, often, the experiences of veterans contrast with these official claims. Certain decisions – such as not providing all men with dosimeters to measure radiation – have led to later concerns and disputes between the British government and nuclear test veterans. Veterans' negative perceptions of the health and safety culture during the tests reveal that their experiences have had a meaningful impact on their mental wellbeing and their perception of risk. While the British government has refused to issue compensation to those who believe their health conditions have been caused by their participation in the tests, other governments, such as those of Australia and New Zealand, have. Many British veterans have criticised their treatment, and some believe that a systematic cover-up has occurred within the British government.

As Britain endeavoured to present itself as a modern and technologically advanced nation, personnel on the island faced basic, harsh conditions and limited resources. The effect of this on the mental and physical health of personnel was significant: outbreaks of dysentery and food poisoning were common, as was sunburn and heatstroke. Morale was low, and several servicemen committed suicide. Poor risk assessment and management led to incidents in which men were exposed to acute dangers or, in several cases, lost their lives. The difficulty and danger of life on the island was intensified by the nuclear tests, and veterans' descriptions of protective clothing, line-up drills and radiation sampling provide a vivid narrative of the realities of nuclear tests and expose how these experiences shaped the day-to-day lives of personnel.

Grapple Slings and Moonshine

The Nuclear Families study collected the experiences of the men who participated in the Grapple series of tests.[27] One cohort member, Jim, was a keen photographer and carefully documented his time there. Jim had decided to join the army immediately after leaving grammar school. He said 'I left home in 1954, when I was 16. I'd lived away ... on camp for a couple of years anyway'. He had completed his military training by the summer of 1957 and had experienced his first tour of Germany, before returning to his home unit in Ripon, Yorkshire. After travelling home for Christmas, he was redeployed to Christmas Island on Boxing Day

1957. Only the Royal Engineers travelled en masse by boat. 'The whole unit went on a specially chartered train to Southampton, all thousand of us, then onto the boat and off we went …'. The young soldiers shared facilities in the hull of ship, sleeping in bunks that were three or four beds high.

> The ship got out of Southampton and then trundled off into Atlantic. Most people were seasick for the first week, until reaching the Bay of Biscay. I was lucky though, as I was on the top bunk and wasn't sea sick. The first week there was no queue for breakfast because of the seasickness. It was great until people started to recover, then you can imagine the queues.

He said, 'There was tombola to keep us entertained and guard duties to keep us busy … We'd practice shooting on deck, throwing floating things over the side to aim at …'. The ship refuelled at Curaçao, then travelled through the Panama Canal, with a stop-off at Panama City. Jim revealed a little about it. 'We had a night out in Panama, we hit the bars and then the strip clubs …'. Hangovers must have been nursed as ship left the docks and set off across the Pacific Ocean to its destination.

Initially, unbeknownst to Jim, this ship also had an additional cargo of veterans' wives and families. He said a little more about this:

> Much of it was a publicity stunt. The boat went out with us all and returned with the soldiers who were currently there. They were able to meet up with their families on the ship, and to have a cruise home together which was paid for by the MOD [Ministry of Defence]. I didn't see many of them, I think we were kept separately.

Whether it was a generous gesture or a cynical publicity stunt, it must have been incredible for these men to have had such an unanticipated reunion. The regiment arrived at Christmas Island in January 1958 after three weeks of travelling. The boat anchored offshore, and soldiers were shipped onto the island by landing craft.

Jim described the soldiers' accommodation on the island:

> The tents were very primitive. We slept on camp beds, with metal legs to clip in, about 6 to 8 inches off the ground. The tents were big, with

at least ten people in each tent. We were provided better beds later ... It was the way things worked, occasionally the supplies arrived before the soldiers.

Large land crabs were abundant on Christmas Island, and 'they crawled into the tent and crawled over you at night'. His solution to this problem was to prop up his bed on jerry cans at night and to hope that he didn't roll out in his sleep. He talked about day-to-day life on camp, and he said that while the main northern camp had washing facilities, the southern camp did not. The soldiers were provided with salt-water lathering soap and instructed to wash in the sea. He said, 'Think of all those naked young men running into the sea, it was quite a sight!' He burst into laughter and jokingly offered photographic evidence.

There was a cinema, a mess and a church. He was proud to say that there were two women from the Women's Royal Voluntary Service who were permanent residents on camp. They had their own accommodation hut and were described by him as 'matronly types'. Jim didn't have much to do with the local population, as the local community lived mainly around the port. He talked about leaving the island; he said that he only left twice during his time there and told me about an unusual example of camp solidarity when he attended a funeral, a burial at sea away from the island. He was required to participate, and I asked him if he knew anything about the deceased. He told me 'I don't know why he died, I just had to do the honours.' No questions were asked.

The second time Jim left the island was more cheerful. All the soldiers were provided with a holiday, and those on Christmas Island were sent to Hawaii. He left Christmas Island for a week of relaxation, but this turned into two when the plane 'conked out'. This plane was also used to pick up supplies, such as fresh fruit and vegetables. When I asked about the food on camp, he described it as 'normal' for the time.

He talked about managing risks to health, and he described the white powdery anti-malaria tablets that were prescribed with each meal. He also told me that the entire camp was 'zapped' daily by an aeroplane spraying DDT to kill mosquitoes. He said it was commonplace at the time, but he was concerned about the effects of inhaling biocide. He also talked about the lack of local travel restrictions on the island. 'I used to go swimming, go walking, I'd borrow a 4×4 and take off to a lagoon in the middle of the island.' He said that he'd learnt to drive without a licence, while working for a scientist from Aldermaston. He would tow and set up the generators for the scientist's cameras. There also were no petrol stations, just jerry cans of petrol which were emptied and left on the side of the road and were later re-filled by another soldier. Unsurprisingly, the soldiers' work and life seemed deeply intertwined during their time on the island.

Jim mentioned that letters were written to the soldiers' families by the commanding officer. He said 'these letters explained where we were and what we were doing, for instance, constructing buildings or repairing roads. My mother kept mine – I discovered them in her home after she passed away. I still have them.' Jim also wrote letters to his mother himself and posted some more unusual souvenirs of his experience. 'I used to post coconut shells back to England – you'd just write the address on the shell'. He shared his final thoughts on the experience, saying 'You can't imagine … it was so exciting.'

Figure 3.1 Young soldiers hanging out on Christmas Island
Source: Photo by R. Watson

However, in the years since working on atomic testing, his perspective has changed. He worries about the health of his contemporaries and the impact upon their families. He has requested his own military medical records with little success. While he is relatively healthy now, he has experienced several health problems during his life that may or may not be attributable to his work on the Grapple tests. He worries about the availability of information for atomic veterans, and wonders whether this material has been intentionally misplaced rather than lost.

Another British nuclear test veteran, Gerald, described his experiences of a military hierarchy that prevented questions being asked. Gerald celebrated his twentieth birthday one week after arrival on the island and his description of this provided an interesting insight into camp social culture. He said that 'there were no spirits available for us lads, only the officers had spirits, so we got drunk on warm tinned Guinness and brandy. The Guinness was out-of-date, and the refrigeration system had broken.' The senior ranks relaxed and drank Tom Collins and Grapple Sling cocktails in the officers' facilities, delicate combinations of gin or whisky with tonic water, lemon and sugar syrup, while the troops were attempting to make their own potent moonshine, brewed in gallon jars covertly buried in the sand beneath their tents, to supplement their 'tinnies'. I asked Gerald how he felt about being a nuclear test veteran, and he told me: 'I felt privileged to see something that no one else had, it was a unique experience.'

Jim's and Gerald's stories of time spent on Christmas Island share parallels, but each also has a uniqueness determined by that man's own personal history and positionality – what is recalled and what is forgotten, what is significant or trivial to each. The experiences of the British nuclear test veterans show naivety towards the consequences of risks on the island, but hopefulness and pride at playing a role in a unique part of history.

This is also reflected by international research of memories and perceptions of exposure to ionising radiation. Garcia et al. studied 16 USA atomic veterans and discovered that resolution of the events that occur during nuclear weapons testing requires emotional and cognitive processing that is in contradiction to prior beliefs.[28] This study found that veterans felt ineffective and helpless in relation to a powerful and unresponsive government; in some cases they felt isolated from other atomic veterans, and experienced self-doubt around their own experiences. This is due to the dramatic narrative change that has occurred as the true nature of nuclear weapons has gradually been revealed.

A Time Bomb?

It is hard to understand the true health, wellbeing, cultural and social implications of nuclear weapons testing. It is difficult to obtain data from all affected regions and nuclear weapons testing participant nations as the topic is still subject to high levels of security and confidentiality. There is an international geography of secrecy. However, studies have been undertaken that try to provide insights into the life outcomes of the atomic veterans.

There are many challenges with research of this nature that link back to the culture of secrecy that used to surround the veterans' work. Some veterans can be reluctant to talk about their experience or to attribute blame for their health challenges to their state military organisation, due to concerns about prosecution. Other atomic veterans are convinced that the bomb is responsible for their life challenges, resulting in conspiracy theories emerging around their experiences. A singular universal challenge for these communities is addressing the lack of information that has been provided to them about their experiences and the risks that they and their families may, or may not, face. This information inequality and lack of support has resulted in elevated perceptions of the risk from ionising radiation exposure within the nuclear community of atomic veterans and their families. This has had repercussions, as perceptions of risk can change life choices.

Very little work has been undertaken recently on the experiences of nuclear veteran families, with a single study of seven atomic veterans and their families in the USA published in 1990 by Murphy et al.[29] This study noted a significant psychological effect on family members. Reasons given for these effects included the invalidation of their experience by government and authority figures, concerns about genetic effects to future generations, a desire to protect each other from fears of physical consequences, and a need to leave a record of their experiences to prevent future suffering. The Nuclear Families study interviewed 144 participants, including 23 UK atomic veteran families, to try and discern attitudes towards risk, experiences of disability, and the community experience of being part of a veteran family. The study discovered challenges associated with caring for older and unwell veterans, heightened perceptions of risk of genetic effects and general health effects. Descendant daughters were electing to not have children because of these perceived risks. There is a lot of concern and fear about ionising radiation and health, with common health conditions being misattributed by veterans' children to their fathers' experiences. It is significant that sociocultural effects are occurring inter-generationally across these communities.

Confounding Health

The US and UK atomic veteran communities, as well as military data pertaining to the tests, have been explored by the scientific, military and academic communities. There are parallels to the experiences of these transatlantic atomic communities, and therefore parallels regarding their health and lived experiences. Experimental epidemiological and cancer studies have provided some evidence of health risks, albeit with uncertainties surrounding radiation doses of 100mSv or less. However, it is very difficult to untangle and identify a single cause for the health challenges that have been experienced by the atomic veterans. There is neither reliable exposure data nor an understanding of what else they may have been exposed to (DDT, asbestos, beryllium, and other carcinogens were common at the time). Therefore any evidence of health problems cannot be attributed conclusively to ionising radiation.

The issue of identifying health consequences is confounded by the life choices that atomic veterans make. Being a member of the military has a protective effect on health, known as the *healthy soldier effect*.[30] Members of the military must go through training, where less healthy individuals leave, creating an elite of healthy people. They lead active lifestyles with healthy food during their working years and, as soldiers, they receive more medical attention and physical inspection than the average population. Most become used to routine medical examinations, and this self-monitoring and other healthy behaviours can persist after leaving the military. This effect confers a 10–25 per cent decrease in risk of mortality for those who have served in the military, compared to the general population. However, the *healthy soldier effect* predominantly refers to physical health and wellbeing, with mental health comparatively neglected. For instance, suicides and homelessness are both notable veteran mortality risks.

Academics have undertaken several studies on the US, New Zealand and UK atomic veteran cohorts to try to understand health effects and risk of health problems. Bross and Bross's reanalysis of the 1985 USA National Research Council report on *Mortality of nuclear weapons test participants*, shows 62 per cent higher incidence of leukaemia and digestive, respiratory and other cancers among soldiers involved in nuclear weapons testing whose reported doses were over 300 mrem.[31]

In the UK, there have been several epidemiological studies of British nuclear test veterans, including three studies by Darby and Muirhead in the 1990s.[32] However, difficulty arises again due to confounding factors, as any mortality or cancer incidence detected may be due to other agents or exposures. It is known that ionising radiation was not the only risk, and that the healthy soldier effect goes some way to offer a protective capacity, so it is very difficult to tease out the true impact of ionising radiation. Muirhead et al. published a study in 2003 that explored the health effects to a total of 21,357 servicemen and civilians, who participated in the tests, and were followed over the period from 1952 to 1998.[33] This group was countered by a control group of 22,333 men who had not participated in nuclear weapons testing. Analyses were conducted for mortality and incidence of 27 types of cancer. It is important to note that this study showed that overall mortality and cancer incidence in UK nuclear weapons test participants have remained similar to those in the control group, who have no suspected exposure to ionising radiation.

Overall mortality has remained lower than expected, compared to national rates. This study showed that there was no risk of multiple myeloma risk among participants. However, there was some evidence of raised risk of leukaemia among test participants relative to controls, particularly in the years immediately after nuclear weapons testing. However, this could be a chance finding, due to unexpectedly low rates among the control group and the generally small radiation doses recorded for test participants. Nonetheless, the possibility that test participation caused a small absolute risk of leukaemia cannot be ruled out. A further study by Muirhead et al. in 2004 demonstrated that there was no evidence of increased risk of multiple myeloma among test veterans in recent years.[34]

In the late 1990s, Roff surveyed the members of the British Nuclear Test Veterans Association (BNTVA) to try to identify health problems, but there are a number of issues with her work.[35] While it provides a good broad insight into the challenges faced by this community, their health problems are self-reported and, because they are all BNTVA members, they are more likely to attribute these problems to ionising radiation. Her statistical study of 1,041 members showed that 84 per cent reported health challenges, including skin conditions, dental problems, a small percentage of participants experiencing cataracts before the age of 40, infertility, early hearing loss and early heavy hair loss. Health issues were also reported among 39 per cent of children and 21 per cent of grandchildren.[36] There is a need for medical records to back up this study, and the self-selection approach means that those who have experienced difficulties with health are more likely to come forward.

A study of Australian veterans of the British nuclear tests again showed again that all-cause mortality was not raised.[37] However, mortality and incidence were raised for cancers of the head and neck, lung, colon, rectum and prostate, and for all cancers combined. For oesophageal cancer, melanoma and leukaemia, incidence was significantly raised but mortality was not significantly raised. Again, our 'healthy soldiers' may have experienced more medical check-ups that meant that their health challenges were managed more effectively. It is worth noting that melanoma is associated with sun damage, which may offer a more plausible explanation for skin cancers among the predominantly Caucasian soldiers working in the Australian desert.[38] Oesophageal cancer is also common among those who drink alcohol and smoke heavily, two lifestyle factors of the era, especially in the military. This study found that there was no association between radiation exposure and overall cancer incidence or mortality, or of any cancer or cancer deaths occurring in excess. Contributing factors included smoking, alcohol and asbestos exposure, and demographic differences to the Australian population with whom rates were compared.[39]

More recent studies in the USA have shown that exposure to low-dose ionising radiation does cause a tiny increase in the risk of leukaemia, although it is debatable as to whether this risk is significant. Clearly further work is needed to understand the issue of health and the atomic veterans. The Million Persons study is currently attempting to resolve some of the lingering questions that remain, including 115,000 atomic veterans in its cohort of people who may have had an occupational radiation exposure.[40] Veterans have been traced through military records to remove the effects of self-selection. The study is trying to discern the risk from gradual exposures over time, rather than brief exposures to elevated doses of ionising radiation. It explores the consequences of internal and external doses of ionising radiation. Its aim is also to estimate the lifetime risk of radiation-induced leukaemia. Perhaps this study will provide some more concrete answers about experiences of health within the veteran community.

Contesting Diagnoses

Just because you cannot see a problem, doesn't mean that it isn't there. Health challenges attributed to environmental exposure are frequently contested. Diagnosis is complex, and our scientific understanding is limited. Sometimes conditions can be unaccepted by the medical community, or *unmedicalised* until it is accepted that there is a problem. Examples of conditions that have historically been unmedicalised include chronic fatigue syndrome, post-traumatic stress disorder, and Gulf War syndrome. However, there are specific difficulties with gaining recognition of an environmental exposure.

Trundle's work has explored the experiences of the nuclear test veterans and determined that they needed to provide three levels of 'proof' to gain state recognition for their illnesses: a *biomedical disease label* from sanctioned medical experts, proof of exposure, and proof of a causal link between exposure and disease.[41] For many nuclear test veterans this is impossible. Illnesses often remain unmedicalised or invisible, there is a lack of records of individual exposure rates, and it is very difficult to prove a link in the form of a scientifically legitimised and politically recognised aetiology. Among nuclear veteran communities, a diagnostic practice is desired that affirms the somatic nature of illness, but also asserts a politically and morally configured notion of culpability.

Atomic veterans often want a medically verified explanation for their illnesses, and actively work to try and remove perceived negative *political influences* from the diagnostic process. This means that they contest some explanations of their diagnoses, for example ascribing their skin cancer to ionising radiation as well as, or instead of solar radiation. While not denying the *biological* nature of their afflictions, the atomic veterans place a significant emphasis upon revealing a *political cause* for their disease – that is, government culpability. This can be described as a quest for a biopolitical endpoint, where historical narratives are included about a nation's shame and a state's admission of guilt. The veterans are demanding an endpoint that enables them to assume the status of a collective who have endured a grave injustice and are therefore perceived to be entitled to public recognition, state resources, a service medal and an apology. However, it has been extensively debated as to whether medals should be awarded for work that is retrospectively recognised as causing environmental and humanitarian harm.

Internationally, atomic veterans claim to suffer multiple health problems from radiation exposure and seek compensation from the state. They contest and devalue military and medical records, and instead they elevate their personal and collective memories, based on what they have witnessed. An example of this is the myth of the 'x-ray hands', where veterans of Christmas Island claim that they saw through their gloves and skin, right through to the bone, due to ionising radiation during the nuclear weapons tests. However, the type and nature of ionising radiation produced during the blasts would not enable such a phenomenon. A more likely explanation is the powerful burst of light produced at the moment of detonation.

Atomic veterans' organisations continue to resist state evidential and archival materials. They accept certain documents as historical truths, but only if they confirm the communities' understanding of the atomic scenario, and emerge from archives without state sanction. Atomic veterans' organisations have therefore created their own private archives, which function as sites of legitimisation for their perspectives, perceived legal proof, and also serve to memorialise other members of their community. Therefore, the atomic veterans' organisations subvert and mimic the documentary logic that already exists within state records. They are reluctant to share their archives with other organisations, which makes it difficult to gain a true understanding of their experiences and perspectives.

NUCLEAR COMMUNITIES

There is considerable support available for atomic veteran communities internationally, and specific schemes have been implemented by USA, French, Australian, New Zealand, Fijian and British governments. Still, these communities campaign for more support. They desire more funding to try and understand their own experiences, and to try to ensure that their descendants receive ongoing support.

In the UK, British nuclear test veterans are supported by a Ministry of Defence (MOD) team that addresses their specific health concerns and helps with their applications to the Armed Forces Compensation Scheme. In addition, further support is provided for initiatives such as Armed Forces Day and for HM Armed Forces veterans' badges. Support has also recently been provided in the form of large-scale government funding for independent research. However, the veterans want a specific formal government medal to recognise their work on the atomic bomb, and the MOD has not been forthcoming with this. An unofficial paramilitary medal that was commissioned by the British and Australian nuclear test veterans' associations can be purchased for £45 online, and many veterans own this medal. The UK has two support and issue campaign groups: the BNTVA and Fission Line, a smaller organisation that campaigns for atomic veteran justice.

In the USA, compensation is available to veterans who have any one of 21 cancers that are traceable to radiation exposure – these men are entitled to a one-time award of up to $75,000 or a monthly disability payment from the Department of Veteran Affairs.[42] However, there are concerns about access to this scheme – it is difficult to verify records of service, and there is no formal discharge form for US atomic veterans. There are also difficulties surrounding the culture of secrecy around US nuclear defence work, which meant that veterans could not discuss their experiences until 1996.[43] However, while the DD-214 discharge form does not mention atomic weapons testing, it is widely known that this signifies work undertaken on atomic weapons. Just like in the UK, those involved in clean-up operations, such as the Enewetak Radiological Clean-up, are in a place of limbo as they have not been formally recognised as atomic veterans in the same way as those men who directly participated in weapons testing, despite it being a radiation risk activity.

American veterans are also supported by the National Association of Atomic Veterans (NAAV), an organisation formed by a group of ex-military personnel who were first-hand participants in the US atomic testing programme that provides solidarity and pursues their cause. NAAV has successfully campaigned for a US government Radiation Dose Reconstruction team to approximate radiation exposure to atomic veterans as part of the Million Person Project. The NAAV has suggested that they want the government to acknowledge that this community was subjected to an unusual risk, beyond usual military service. They also feel that the government should provide them with free medical care for conditions that may be due to exposure to ionising radiation. Similarly to the UK, some in the community feel that they should receive financial compensation for their work, whereas others want no more than official recognition, by way of a certificate or medal.

Internationally, there is a hidden community that has been affected by the nuclear tests and is not always supported: the wives and children of atomic veterans. This community often feels at risk from the ionising radiation their family members may have encountered. This community also experiences the hidden challenges of caring for and supporting aged veterans. There is great concern within atomic veteran communities that their role in the atomic bomb will have genetic effects, adversely impacting their children. However, any spermatozoa affected by the radiation will have been naturally replenished before they could cause any genetic defects, so long-term genetic health effects are extremely unlikely. However, the concern and anxiety caused by this perception of risk is undeniable. Moreover, the paucity of information provided to these communities has resulted in an amplified perception of these risks. Some daughters of atomic veterans have decided not to have children due to perceived concerns about damage to their own DNA.

Many of the concerns stem from events that are harrowing, but also common to the normal population. A US study exploring reproductive outcomes for veterans shows that adverse reproductive outcomes are not as rare as one might think in the general population. This includes the inability to conceive, the premature spontaneous termination of a pregnancy, the birth of infants with a congenital malformation, and premature death. The study estimated that 15,000 children with major birth defects would be expected among the 500,000 or so offspring of the 210,000 Atomic Veterans, even in the absence of any radiation effects. This is important, as it quantifies the actuality against the perceived risk, and demonstrates that the community is within the bounds of normality in this sense.[44] It suggests that there is an 'atomic veterans syndrome' rather than a specific and easily diagnosable physical health problem.[45]

Historic self-reported health studies showed that one in seven of British atomic veterans in a sample of 1,014 did not father any children after they returned from testing, equivalent to 14 per cent. This is also the average rate of infertility for men and women, and the likelihood of male infertility increases with age.[46] There are also other host and environmental factors that influence descendant health outcomes, including maternally or paternally derived inherited defects, exposure to smoking and the consumption of alcohol during gestation, pre-existing maternal illnesses such as diabetes or other illnesses during pregnancy, and poor nutrition. During the 1950s, 1960s and 1970s, these factors would have been more prevalent, as public health measures such as five-a-day for diet and anti-smoking bans were yet to come into place. It is therefore almost impossible to have an epidemiologically valid study of descendant effects and health outcomes for any person, within the domain of low-level ionising radiation exposure.

A study also reported that nearly half of the health problems among the 5,000 studied offspring of the nuclear weapons test veterans consist of the same dermatological, musculoskeletal and gastrointestinal conditions that their fathers have also suffered from.[47] This is likely to be an inherent hereditary, rather than ionising radiation exposure related link, as many of the conditions reported in this study are common hereditary complaints: eczema, dermatitis and rheumatoid arthritis. Although this provides interesting insights into the experiences of health within the cohort, the reported rates of descendant health conditions are not significantly different to those reported by the general population.

It can be difficult to communicate the reality about the health risks to veterans and their families. Experts and the state are distrusted, and the veteran communities tend to understand their perception of risk through personal, relational and affective experiences. Risk likelihood among the community is identified based on the misfortune of those that they know, rather than on statistical realities. Understandably, the negative experiences of other veterans capture experiential knowledge, and reveal personal and familial suffering in accessible and relatable ways. Unfortunately, expert evidence from specialists in radiation protection or biomedicine has not convinced this community. The atomic veterans use their experiences to build narratives about heritable and ionising radiation related illnesses. This has led to high levels of anxiety around health and the influence of genetic heritage being reported by the families of British atomic veterans to the Nuclear Families project. This also leaves this community susceptible to exploitation by fringe academics and experts, who provide 'radiation gene tests' and ineffective medicines, fuelling their concerns.

REMEMBERING THE BOMB

Atomic veteran communities worldwide have all found ways to memorialise their experiences. The militarism of their time testing the bomb means that, for many, rituals involving public parades, official military or military-like activities, provide an opportunity to commemorate and memorialise. The BNTVA has regular trips away and events for its community that provide cohesion, solidarity and support for the community and their families. These activities are mirrored by French,

US and Australian veterans support groups internationally. In 2016, 2017 and 2018, for example, British and French atomic veteran groups joined in solidarity to march together to the Arc de Triomphe to rekindle the perpetual flame above the Tomb of the Unknown Soldier.[48] In 2018, the BNTVA also revisited 'Christmas Island' (now Kiritimati), to commemorate 60 years passing since the first hydrogen bomb test.

Figure 3.2 The chairmen of the French and British atomic veterans' associations rekindling the flame under the Arc de Triomphe, Paris
Source: Photo by B. Alexis-Martin, 2016

The BNTVA community has become 'big business' for its members, who sell memorabilia and trinkets such as jackets, ties and badges with the BNTVA emblem online and at events. It communicates with its community through regular newsletters, and also has an official Twitter and Facebook page. As an aged veterans group, it is known that loneliness is common within this community and the BNTVA has an important role in bringing these people together. Similarly to veterans in the US, France and New Zealand, the BNTVA has campaigned for veteran justice, bringing veteran issues to parliament and petitioning for independent studies of their community. The BNTVA also provides memorialisation services, such as coffin drapes, and run an *In Memoriam* webpage for their deceased members. Compared to other communities who have been affected by the bomb, such as indigenous communities worldwide, the atomic veterans have received a huge amount of support and research funding for their plight. It is time to decolonise the bomb and give more support to the women, children and families of the atomic veterans, and the indigenous communities who have been affected by nuclear weapons testing worldwide.

4
After Nuclear Imperialism

> Nuclear weapons ... are the ultimate coloniser. Whiter than any white man that ever lived. The very heart of whiteness.
> – Arundhati Roy, *The End of Imagination*

Nuclear weapons testing has had grave consequences for military personnel and the scientific community. However, these effects have come to be quietly overshadowed by the severity of outcome for indigenous communities, the largely unknown people who originally lived in the places where nuclear weapons were tested. The careless and thoughtless exploitation of these people has been shrouded by official histories. Local victims of the atomic test series were depicted as less-than-human savages, and they were denied their rights to their homelands in the name of military progress. The colonialism and casual racism of nuclear weapon developer states stripped indigenous people of their dignity and plunged them into precarity and danger. Their homes became the testing grounds of the nuclear military industrial complex. These local communities had no choice, control or autonomy over a geopolitically driven matter of life or death. International Physicians for the Prevention of Nuclear War (IPPNW) estimate that 'roughly 2.4 million people will eventually die because of the atmospheric nuclear tests conducted between 1945 and 1980'.[1]

Disregard for local residents was a universal feature of the planning process for nuclear weapons testing internationally, as 'empty' spaces were sought out and dominated. *Nuclear colonialism* is the 'taking and destruction of other people's lands, natural resources, and wellbeing, for one's own benefit, in the furtherance of nuclear development'.[2] During the Cold War, these last outposts were portrayed as desolate 'wastelands', described as uninhabitable by the military and scientific establishment.[3] The military were then able to territorialise these spaces and make them their own, without opposition. This space invasion forcibly drove many local civilians from their homes, to become *nuclear refugees*. The repercussions of this nuclear colonialism included trauma, health challenges and the loss of cultural ties as communities tried to adapt to their unsuitable new conditions. Some people, such as the Aboriginal Australians of Maralinga, tried to return home, only to discover barbed wire fences surrounding their ancestral homelands. Other communities, including the Kazakh nomads of Semipalatinsk, were left in place to bear the brunt of nuclear weapons testing, and now live with inter-generational genetic health effects. The 167 residents of Bikini Atoll were forced to relocate to another atoll, with Bikini left polluted for the foreseeable future.[4]

After these places were destroyed by nuclear weapons testing, some attempts were made to restore and remediate radioactive fallout and other toxic pollution in the years that followed. However, a perception of inhabitability remained in the state mind-set. This meant that some testing locations became repositories for nuclear waste, such as Runit Dome on Entewak Atoll in the South Pacific. Other places needed repeated remediation and restoration, such as Bikini Atoll, which was bombed 23 times over twelve years.

In recent years, some of these toxic landscapes have been returned to their original communities. However, the health and environmental impacts have already gone beyond the boundaries of people's homelands. For those who were not displaced from their homes, the economic challenges and social stigma of living in a contaminated environment means that relocation elsewhere is almost impossible. Nuclear weapons testing sites are contaminated, but they have still become sites of historic and existential tourism, where the public can become legitimised voyeurs of a 'shocking atomic legacy'.[5] Nuclear tourism can provide a source of revenue for indigenous communities, but they also give a revisionist and sensationalist history of the impacts of nuclear weapons. The authentic experiences of the indigenous people who continue to be affected by nuclear weapons testing is both neglected and tragic.

The global reach of the nuclear weapons tests and their influence on local communities cannot be underestimated. Atmospheric nuclear weapons testing continued across several decades following the Second World War, despite universal knowledge of its harmful health and environmental effects. The US, Soviet Union, UK, France and China tested predominantly across 16 sites located in 9 different countries, across 5 continents, displacing and affecting a vast multitude of different communities, cultures and societies.[6]

The impacts on local people should not be underestimated. Colonised but inhabited atolls and islands of the South Pacific Front were subject to the largest number of French, British and American nuclear tests.[7] The US conducted tests in Nevada that displaced Native American communities.[8] South Pacific Islanders of the Bikini and Eniwetok Atolls in the Marshall Islands, Kiritimati Island, Johnston Island and Kalama Atoll were also displaced by 106 tests undertaken by the USA.[9] France displaced the nomadic communities of Reggane in the French-Algerian Sahara, and the islanders of Mururoa and Fangataufa in French Polynesia during their testing regime.[10] The UK tests displaced the nomadic Aboriginal communities of Maralinga and Emu Field in South Australia and the original residents of Kiritimati Island, which they called 'Christmas Island'.[11] The Soviet Union displaced the indigenous communities of Novaya Zemyla and Kapustin Yar in Siberia; and also affected locals in Kazakhstan, Uzbekistan, Turkmenistan and Ukraine.[12] China displaced the native communities in Lop Nor, with their last nuclear weapon test undertaken just before the Comprehensive Test Ban Treaty in 1996.[13] Each test had different characteristics and consequences for the surrounding community, due to differences in the frequency and scale of exposure, and the local density and demography of the population.

BLACK MIST

The first British atmospheric nuclear weapons tests were undertaken in Maralinga, Emu Field and Montebello Island in Australia in the 1950s.[14] Both the British and Australian governments described Maralinga and Emu Field as a 'wasteland' that was not 'valuable', a place that did not matter in the scheme of things.[15] The indigenous Aboriginal Australians who lived there were barely considered to be human.[16] These communities were exposed, displaced and humiliated. They were subject to land confiscation, cultural desecration and harmful health effects – from ionising radiation exposure, but also due to the psychosocial consequences of cultural dislocation.

The tests at Maralinga and Emu Field had consequences for both Aboriginal and colonial local communities, as approximately 8,590 Australian civilians also participated in the tests. However, no one knows the true number of indigenous people who were affected; since they were not counted as citizens and were not included on any census until 1967.[17] Maralinga was inhabited by the Pitjantjatjara and Yankunytjatjara people long before the British occupation of Australia, centuries before nuclear weapons testing began.[18] However, these indigenous communities were not recognised or protected by the Australian state.

Maralinga means 'thunder' in the Yolngu language, and the place where the bombs were tested had long held spiritual significance for Aboriginal communities.[19] The connection of Aboriginal communities of the Yalata region to the land and their culture of environmental protection goes back thousands of years.[20] They have retained a distinctive cultural identity that spiritually and physically binds them to the natural environment and oceans of Australia. This closeness to nature is seen in their distinctive practices, customs and artworks. The sacred objects of the Yalata community have been relocated, but the enforced displacement of indigenous populations has prevented traditional ways of life being passed on to future generations.

Attempts were made to relocate Aboriginal communities to a new settlement at Yalata and to prevent any further access to Maralinga but many continued to move throughout the region at the time of the tests.[21] For some reason, it was difficult for the British soldiers to communicate the risks of nuclear weapons test sites to the Aboriginal communities that returned.[22] It was later discovered that a traditional Aboriginal route crossed through the Maralinga testing range. This denial of their heritage caused tremendous social and cultural challenges for the Aboriginal community.

It is also known that some of the British Australian tests, such as Totem I, took place under conditions that produced unacceptable levels of fallout.[23] It is almost impossible to ascertain the total scope of health effects to the Aboriginal people, and to discern their causes. This is due to poor health care support, low levels of reporting, the effects of displacement and migration, and the pervasive discriminatory attitude towards Aboriginal people.[24] The traditional Aboriginal culture and lifestyle is open-air. Consumption of local products and water, and the lack of amenities such as piped water, permanent dwellings and drainage facilities would have increased their vulnerability to exposure. Aboriginal people could have received a radiation dose over five times greater than that permissible for members of the public or armed forces.[25]

The radioactive fallout was called '*puyu*' or black mist, by the Aboriginal people.[26] Mima Smart, an Aboriginal community leader, said that 'A lot of people got sick and died.'[27] A report by J.K. Symonds, formerly of the Australian Atomic Energy Commission, suggested that the Australian government knew that British experiments in the late 1950s contravened a moratorium on nuclear tests agreed in October 1958 at Geneva.[28] The McClelland Royal Commission records instances of barefoot Aboriginal people walking across contaminated ground and camping in radioactive craters and scenarios of radioactive cloud drift.[29] Monitoring the movements of Aboriginal people in the region was deemed unnecessary due to racist preconceptions; as Air Vice-Marshall Menaul said, they 'sleep most afternoons'.[30]

Officially, this indigenous community remained legal and physical absentees. After experiencing health challenges, this community have also suffered stress and anxiety about their long-term fate. Dislocated from their homelands and traditional culture, and unsupported by Australian society, unemployment, alcoholism, suicide and pervasive poverty have become notable challenges for the remaining Pitjanjatjara people.[31] The tests have left a legacy of distrust of the Australian government by the Aboriginal community.

The success of Aboriginal homeland restoration attempts is debatable. There have been two phases of remediation since nuclear tests at Maralinga were concluded in 1963. Operation Brumby took place in 1967, and has been described by Professor John Keane as a 'quick brooms-around-the-toilet-floor effort by British army engineers'.[32] In 1980, surveys conducted by the Australian Radiation Laboratory revealed that the site was still not safe, prompting a further decontamination operation.[33] This second phase of remediation was completed in 2000 and has been subject to criticism.[34] Regardless, the Australian government claims that the remediation project was successful and that most of the test site is now safe to visit, excluding a 120 square km exclusion zone (approximately the size of Manchester, UK) that will remain contaminated for the foreseeable future.

In 2014, the land that was taken from the Aboriginal people of Maralinga was finally returned. Nonetheless, the legacy of the tests is ever present. ' I don't want to go back,' Mima Smart told BBC journalist Jon Donnison in 2014, 'Too many bad memories.' Dr Archie Barton, a member of the Tjarutja people, said 'I just want back my mother. I want back my land too. Clean.'[35] However, access has been restricted. Maralinga-Tjarutja general manager Richard Preece saw opportunities, declaring that 'We are going to set up bus tours so that people can be taken round by Robin [the Australian tour guide], who is a walking encyclopaedia of Maralinga'.[36] The site has been set up in trust for the

Aboriginal community by Maralinga Tours, and offers the interested public an opportunity to visit the tour site and gawp at the residual legacy of the atrocities that were committed here. Little compensation has directly reached the affected Aboriginal communities, with the British Supreme Court declaring that requests for compensation have come too late for victims to prove that their health challenges originate from exposure to ionising radiation in the 1950s and 1960s.[37]

NUCLEAR REFUGEES

The indigenous communities of the Pacific Ocean were subject to British, American and French nuclear weapons testing. This was undertaken across the Marshall Islands, the Central Pacific Islands and French Polynesia. These tests have created significant humanitarian concerns. Some indigenous people of these islands have been permanently and repeatedly displaced, experiencing loss of land and of islander cultural heritage. The health, socioeconomic, environmental and cultural consequences of the tests remain to this day.

The US undertook 106 nuclear weapon tests across its Pacific territories, beginning with the Marshall Islands in 1946.[38] This is despite the US having been tasked by the United Nations (UN) to support the self-determination of the Marshall Islanders after the Second World War, as part of the former Micronesian colony of Japan. Instead, the US saw an opportunity for its own military self-determination, and a new place to test the bomb. As a new American colony, the Marshall Islands became the primary location where the United States would test increasingly destructive nuclear weapons and delivery systems.[39] By sanctioning US imperial control over Micronesian lands, the UN Trusteeship Agreement hastened the rise of an imperialistic military industrial complex during the Cold War.

A 2012 report of the UN Special Rapporteur found that 'nuclear testing resulted in both immediate and continuing effects on the human rights of the Marshallese' including loss of livelihood, land and indefinite displacement. US military officials had described the region as 'sparsely populated', when, in fact, hundreds of Marshallese people lived there, subsisting through sustainable agriculture and fishing. The Marshallese Islanders were ordered to leave their homes 'for the good of mankind and to end all wars'.[40]

Lemyo Abon, who lived on Rongelap Atoll, in the Marshall Islands, said: 'For almost 60 years, we have been displaced from our homeland, like a coconut floating in the sea with no place to call home.' Only part of Rongelap Atoll has been remediated since the tests and many islanders have never returned home. The population of Bikini Atoll were also repeatedly displaced.[41] This community was initially relocated to Rongerik Atoll in the Marshall Islands, a desert island that lacked the resources to support this new population. People began to starve and were relocated to Kwajalein Atoll and Kili Island. However, there was no lagoon environment there to sustain their traditional way of life.

These atolls were converted into neocolonial outposts by the USA. To tempt the families of scientists and engineers there, housing on Kwajalein Atoll was modelled on the picket fences and landscaped gardens of Levittown, New York. It was transformed into a space of suburban domesticity, a piece of 1950s America in the middle of the Pacific Ocean.[42] In 1949, a local community of Micronesian and Marshallese labourers were excluded from Kwajalein Atoll, when it was claimed that suitable living space could no longer be provided for them on the island. This workforce was invited back to undertake menial jobs on the atoll but could not live there. Instead, they were subject to curfew and returned to nearby Ebeye Island each evening by a police-escorted ferry. By 1951, the atoll had a solely American population, while the people of Kwajalein Atoll became foreign commuters to their own lands, entering colonial America each morning and returning to Micronesia each night.

It was accepted by American civilians and the military that this was appropriate, with island guides for incoming American employees detailing the availability and cost of Marshallese domestic help, noting that: 'They are transported to Kwajalein atoll in the morning and returned to their island at the end of the day.'[43] Other guides described how 'maids are also given physical examinations at the hospital before being employed', introducing the possibility that some Marshallese could be carriers of disease. The UN Institute for Disarmament Research reports that Marshallese women also experienced 'humiliating examinations by US military medical and scientific personnel' as part of the USA health monitoring programme.[44] The 1961 Bell guide included descriptions of Micronesian racial characteristics that included anthropological species language such as 'breed lines', in an article entitled 'Basic Micronesian racial stock'.

Nuclear testing ended on the Marshall Islands in 1958, after 67 airburst tests. There were some limited opportunities for islanders to return home. The US originally declared Bikini Atoll safe for resettlement and some residents could return in the 1970s.[45] However, they were displaced yet again in 1978, after accumulating an elevated dose of radiation from consuming local produce on the atoll. Similarly, those returning to Rongelap Atoll in 1957 fled the island again in 1985, amid correct fears of residual radiation.[46] Just one of the 60 islands in the atoll has been cleaned up by the USA, at a cost of £30 million. The humanitarian and civil rights issues that have been created by the US Pacific nuclear tests have been gradually coming to light, and environmental and health issues remain. As with the British test sites in Australia, clean-up operations have been limited and contamination is scattered across the Pacific testing grounds. According to the IAEA (International Atomic Energy Agency), the permanent rehabilitation of Bikini Atoll is no longer appropriate, due to the severity of radiological contamination, where remedial actions such as removal of topsoil could cause serious environmental harm.[47]

Radioactive waste is still stored on the Marshall Islands in concrete cascades, such as Runit radioactive repository dome on Entewak Atoll.[48] Sea level rise and climate change now put these places at risk. Despite an eight-year remediation operation after testing ceased, the US government has refused to fund a comprehensive decontamination programme that would make the entire atoll fit for human resettlement again. In 2016, the Marshall Islands took India, Pakistan and the UK to court in a landmark case for failing to halt the nuclear arms race. It was thrown out by the International Court of Justice.[49]

The British Pacific tests were undertaken on Malden Island and Kiritimati Island in Kiribati Atoll. In 1956, the year after the UK announced that nuclear testing would proceed on Kiritimati Island, Western Samoa petitioned the UN Trusteeship Council to halt the tests, and the Rarotonga Island Council expressed concern and asked that the testing area 'be situated at some greater distance'.[50] The local communities were either ignored or paid to undertake menial labour for the military. In 2015, Kiribati's Permanent Representative to the UN, Ambassador Makurita Baaro stated, 'Today, our communities still suffer from the long-term impacts of the tests.'[51]

The British Grapple tests on Christmas Island considered the likelihood of exposure to local communities with racist terminology. A

UK report stated that: 'for civilised populations, assumed to wear boots and clothing, and to wash ... It is assumed that in the possible regions of fall-out at Grapple there may be scantily-clad people in boats to whom the criteria of primitive peoples should apply.'[52] Further talks in 1956 with the UK defence minister suggested that only a 'very slight health hazard would arise, and that only to primitive peoples'.[53] Local communities were not impressed by British health and safety culture, with Indo-Fijian newspaper *Jagriti* noting that 'Nations engaged in testing these bombs in the Pacific should realise the value of the lives of the people settled in this part of the world. They too are human beings, not "guinea pigs".'[54]

Early French nuclear weapons testing took place in the Algerian Sahara. These original Saharan tests resulted in the contamination of desert sand, and the Algerian government has claimed that radiation is 20 times higher than expected in some places near the test sites, with risk of inhalation and ingestion of contaminated particles as desert dust. Anti-nuclear protests on the islands of French Polynesia pre-dated French plans to transfer its tests to the South Pacific. Regardless, 179 tests were undertaken in the Moruroa Atoll and 14 in the Fangataufa Atoll.[55] These tests have increased the incidence of thyroid cancer in the local population, and indigenous workers who were employed in remediation activities received less protection than French scientists. Teraivetea Raymond Taha, a former Moruroa worker who later suffered from leukaemia recalled:

> They were all dressed in special outfits with gloves and a mask. We Maohi workers were just following on behind them, without any special gear to protect us ... The bosses said: 'It's OK, you can go over there.' We were scared, but if we'd refused, we would have been on the next plane back to Tahiti. We would have lost our job, so we went ahead cleaning up without asking any questions.[56]

Indigenous communities across the Pacific Islands are still living with the psychosocial, economic and environmental after-effects of 50 years of nuclear weapons testing. The atomic veterans of atmospheric tests at Maralinga, Emu Field and the Monte Bello Islands have campaigned for enhancements to their own rights from the Australian and British governments, but it has been much harder for the voices of indigenous communities to be heard. Despite a legacy of exploitation, their health concerns have been dismissed and they have not been culturally, socially or economically supported. Fijian soldiers and sailors have gained compensation from their own government, and the Marshall Islanders have lodged multiple petitions to US Congress to increase the level of compensation provided to them for damage to people and property caused by the nuclear tests. The Marshall Islands Nuclear Claims Tribunal awarded personal injury and damage claims arising from the tests but stopped paying after the compensation fund was exhausted.[57] The president of the Marshall Islands, Christopher Loeak, called for the US to resolve unfinished business, as the compensation provided by Washington did not provide a 'fair and just settlement for the damage caused'.[58]

'RADIOACTIVE MUTANTS'

The Soviet Union undertook 715 nuclear weapons tests from 1949 to 1999 on a remote northern archipelago of Russia named Novaya Zemlya, and on what are now the countries of Kazakhstan, Uzbekistan and Turkmenistan.[59] Despite remote locations being selected, people were not always moved from nuclear weapon test sites or provided with protective information. Health and displacement issues have arisen for indigenous nuclear test survivors of the Soviet nuclear weapons test regime, similar to those of the victims of Western nuclear powers. In 1961, for example, the USSR wanted to test the most the most powerful nuclear weapon ever developed, the Tsar Bomba; 500 indigenous people were removed from Novaya Zemlya to create the test site.[60] Fallout from Tsar Bomba travelled across the globe on atmospheric currents, causing elevated levels of ionising radiation to be detected in Norway, Sweden, Canada and Alaska.

The most significant sociocultural legacy of the Soviet nuclear weapon tests exists in Kazakhstan. Semipalatinsk Nuclear Test Site (STS), or 'The Polygon', is nestled in the remote Kazakh steppes. A million people lived within 160 km of the test site, with several thousand people living in villages bordering the STS. These communities were left in place during the 40-year period when 456 individual nuclear weapons tests took place, from 1949 to 1989.[61] Of these, 116 of tests were above ground. It has been suggested that, rather than negligence, this was an intentional decision by the state to better understand how radiation affects the human body. The impact of the tests, the number of victims and the way they have been affected have been disputed, and the secrecy of the test site means that the number of victims may never be known.

The Semipalatinsk region is still heavily contaminated in places. It is inhabited by communities who have gradually become aware of the contamination of their land, but who are too poor to relocate.[62] They are fated to live their lives in this toxic environment. Some members of the Semipalatinsk community have reclaimed the stigma associated with their fate, describing themselves as 'radioactive mutants' and jokingly claiming that they now need the radiation of their beleaguered homelands to survive.[63] There have been significant immediate, long-term and hereditary reproductive health effects for these communities because of radiation exposure. The community struggles with these health challenges, and life expectancy for the exposed community was several years lower than that of a Kazakhstan control group.

Ten villages near the STS test site in Kazakhstan were studied by Japanese researchers in 2002–2004. This project discovered that many villagers had experienced sub-acute radiation injuries at the time of the blasts. From the 1960s to early 1980s, childhood leukaemia had an increased incidence, and, among women of reproductive age who were exposed to radiation during childhood, malformations among their children were reported more often than in control groups. After 1985, the frequency of congenital malformations dropped, and infant mortality decreased to pre-test levels. This is likely to be due to the new generation of women giving birth at this time having less radiation exposure from airburst tests. In a heart health study of this community, however, cardiovascular diseases were reported more frequently and occurred earlier, compared to the control group.[64] Institute of Radiation Safety and Ecology (IRSE) research has shown a spike of cancer incidence in the late 1980s among the exposed population, three times higher than the level of the control group in Kazakhstan. This increase included more deaths from lung and breast cancer.[65]

STS was abandoned by the Russian Ministry of Defence in 1993.[66] The site became a both a health risk and an international security risk. It was left unguarded and neglected during Kazakhstan's post-communist economic crisis, with no state authority exerting any control over this large space.[67] No radiation protection or access restriction measures were put in place, so this dangerous space was now easily accessible to local communities. The sites were looted for any materials that could be sold as scrap. The looters lived by their sources to protect them, drinking the water that came up from test tunnels. Presumably, these people received substantial doses of radiation, but studies by IRSE reveal that registration and public health measures were not conducted to find this out.

Considerable work was undertaken during the late 1990s to reduce risk from unsecured nuclear materials, in conjunction with the USA and Russia. Work began in 1996 and was completed in 2012, when most nuclear materials were placed in Delegen mountain massif repository on the original Polygon site.[68] Visitors to the site are surprised at the lack of barriers and warning signs. Borders that were previously determined by Soviet generals in the 1940s have vanished, and much of the site has become re-wilded. Economic activity has resumed, including mining, agriculture and tourism.

However, the psychological consequences for the local community are ongoing. The local villages feel that they were not informed about the radioactive nature of the nuclear weapons tests, and that they were not warned about the dangers of radioactivity. The community was encouraged to leave their homes to watch the tests, and now feel like guinea pigs of a harmful state experiment. The local communities are predominantly concerned about the potential for continued health impacts to their bodies and those of their children.[69] Escalating this uncertainty is the invisible nature of ionising radiation, and the fact that the effects of radiation exposure can take time to manifest. Radiation is perceived as an 'uncontrollable risk that one is unable to detect without specialised scientific equipment'.[70] The community has not been supported with education in this area. Concerningly, this means that villagers tend to attribute any illness obtained in the village, from an upset stomach to a brain tumour, to nuclear testing and radiation exposure. This psychological stress has a real and negative impact upon life quality. Long after the tests have ended, local communities remain concerned about the potential effects of radiation upon their lives.

CHINA AND SOUTH EAST ASIA

The Chinese nuclear weapons programme began with uranium mining in Tibet, at great human cost. Lop Nor is still cloaked in secrecy, despite being the largest nuclear weapons testing site in the world at 100,000 square km. The test site was established in October 1959, with guidance from Soviet Russia, and detonated its first bomb, known as '596', in 1964. It is known that there have been 23 atmospheric tests since then.[71] At the time of its first test, China was considered a high-risk state. The country was in turmoil, with Chairman Mao preparing to instigate the Cultural Revolution and supporting the North Vietnamese Army in their war against the US. However, these tensions were diffused by the diplomatic efforts of Nixon and Kissinger in the 1970s, in conjunction with Premier Zhou Enlai and Chairman Mao Zedong, so China became an accepted member of the nuclear club. Although there is a good understanding of the technical aspects of the Chinese tests, it is difficult to ascertain the local community impacts.

Lop Nor is a salt desert, remaining from the slow evaporation of an inland sea. For China, the region is Xianjiang province, but the indigenous Uyghur community call it East Turkestan. The province is predominantly populated by the minority Uyghur (or 'Uighur') community and includes the Xinjiang Uyghur Autonomous Region. However, this region was an experimental field for warfare and the people living around the test site are still struggling with the effects of nuclear weapons testing. It is thought that, like indigenous Kazakh communities, the Uyghur people were not warned before testing took place, and the sites were not fenced off to the public.[72]

The Uyghur are a forgotten people. They are one of the most oppressed indigenous communities in the world. They have been denied their basic human rights and have been subject to persecution since China's occupation of their land.[73] Their own culture, language, Muslim religion, and traditions have been suppressed, and generic atheist Han Chinese culture has been forced upon them instead. They have been subject to a practice of systematic exclusion by the Chinese government. In what is essentially a cultural genocide, even peaceful expressions of cultural identity have been inhibited, with 're-education centres' existing to constrain Uyghurs' cultural practices and free will. It is notable that members of the Uyghur diaspora have relocated to Kazakhstan, Kyrgyzstan and Uzbekistan, other places that have been targeted by nuclear weapons testing. A Uyghur man now living in Hiroshima said that:

> In around 1989 and 1990, once or twice a year the sky darkened and pillars of sand and smoke grew, making me realise that they were conducting nuclear tests ... I went to the atomic bombing museum and learned for the first time that they could have an impact on human bodies.[74]

Prior to 1981, the fallout from surface tests was thought to be a contributor to the incidence of cancer and birth defects. However, it is difficult to ascertain if these effects are due to inherited genetic mutations or exposure to a contaminated environment. In 1998, a Channel 4 documentary called *Death on the Silk Road* had a team of doctors and filmmakers pose as tourists to try and assess the effects of the tests on the local population. They discovered that there were many infants with birth defects in the villages. Among the Uyghur community, many were suffering from lymphoid leukaemia. The incidence of cancer in cities in Xianjiang province is 30 per cent higher than the average rate across China, although the Chinese government refuses to acknowledge any ongoing harm to citizens.

These issues are compounded by poor access to medical care, social services and necessary information about radiation health effects. Dr Takada Jun, an expert in radiation protection and representative of the Japanese Radiation Protection Centre, said that the Chinese regime has never allowed any form of independent or outside analysis, evaluation, or study of adverse effects to human health because of the tests. This means that specialised health care is not given to those suffering from radiation-induced health problems or birth defects. Uygur traditional medicine is derived from Unani practices, and not adapted to treat health challenges potentially arising from radiation exposure.

Official compensation for civilian victims is not forthcoming. Only former military personnel deployed to the area during testing, known as the '8023 Force', have received compensation from the state for their and their families' serious health problems.[75] Currently, major state investment in industrial and energy projects means that Uyghur communities are worried that the Han Chinese are taking their jobs, and that their farmland is being confiscated for redevelopment. Chinese colonial and apartheid rule of Xianjiang province is increasingly rejected by the Uyghur people. Uyghur separatists have attacked Chinese targets in recent years and are now subject to a Chinese equivalent of the 'war on terror'.[76] However, most Uyghurs are hoping for a peaceful change.

Globally, the experiences of indigenous people are difficult to quantify and remain poorly understood. These communities were systematically disenfranchised and thoughtlessly cast aside in the name of military industrial development. International treaties relating to nuclear weapons testing currently contain little provision for supporting victims, and nuclear weapon possessor states have been slow to admit blame, instigate reconciliation or provide compensation to these communities. They are internationally linked by the *necropolitical* similarity of their experiences, whereby the state creates a harmful hierarchy of who can live and who must die, often for political means. Colonisers decided that their safety and wellbeing was less important than the state's opportunity to test nuclear weapons, out of sight of anyone who might have had the power or influence to stop them.

Uncertainties about the long-term effects of the nuclear tests make psychological stress and fear an important and continuing legacy of nuclear weapons testing. There is a dire need for recognition, accountability, monitoring, care, compensation and remediation across every site of nuclear weapons testing. The interconnectivity between racism, colonialism and the nuclear era has undermined the dignity of millions of indigenous people internationally. It is too late for the many who have already died, but there is time to create effective survivor networks that can provide the support that these communities are owed, and to help these communities share their atrocious experiences with the world.

5
After Nuclear War

> Only a catastrophe gets our attention. We want them, we depend on them. As long as they happen somewhere else.
>
> – Don DeLillo, *White Noise*

Nuclear landscapes are both created and destroyed. Evidence of nuclear activity becomes hidden by remediation, re-wilding, or by simple neglect. This destruction is not always total; the architecture of nuclear warfare is often too substantial to disappear entirely, or relics could be intentionally left behind to commemorate or memorialise nuclear activities. The cultural memory of nuclear risk can remain traced across the landscape, creating a taboo around land-use and deterring people from spending time there. Alternatively, it may encourage intrepid tourists and visitors to explore, undertaking what is known as thanatourism or 'dark tourism', but more specifically, nuclear tourism.[1] The problem with this is that while it can memorialise and provide insightful understanding of the legacy of harmful historic events, it can also obscure the experiences of those who still remain in nuclear places and spaces, freezing the site as an atomic plaything and ignoring the ongoing nature of nuclear warfare. In particular, abandoned nuclear places and spaces are interesting because they provide a materiality, a physicality, to the way that nuclear warfare affects the cultural, political and physical landscape.[2] Their abandonment, their strange architectural relics and cratered pock-marks, distinguish them as places of (un)natural change. From Maralinga, Australia to the US Trinity test site, these are sites of nuclear pilgrimage, as both pacifist and militaristic visitors arrive and gaze at the vacant vast landscape, while imagining the bomb.

Sites of nuclear tourism are not just abandoned wastelands, and some are distinctly lively. They include nuclear bunkers that have been turned into museums, such as Hack Green Bunker and the Churchill War Rooms in the UK. There are historic exhibitions in decommissioned submarines, including the Royal Navy Submarine Museum. There are atomic heritage museums dotted across the militarised nuclear sites of America, from the National Museum of Nuclear Science & History in New Mexico to Los Alamos' Bradbury Museum and the Smithsonian's Atomic Weapons Museum. Many of these sites present a carefully curated perspective of nuclear warfare that glorifies and glamorises the lives of nuclear industry workers or the process of civil defence.

The materialities of nuclear warfare provide tangible and embodied evidence of the cultural consequences of nuclear activity through 'the concrete of bunkers, in the radio towers, the food stores, the dispersed centres of government'; but also through the processes and structures that surround the nuclear military industrial complex in its entirety.[3,4] Also of interest is the idea of relational materiality, which provides explanations for why and how nuclear weapons exist in their present form, and what this means in both material and conceptual terms.[5]

At a base level, the materialities of nuclear warfare consist of concrete and steel, plutonium and beryllium; in barracks and bunkers; missiles and silos. The physicality of these artefacts of warfare are of interest to the cultural or creative geographer, from the perspective of the purposeful representation of their form and structure. These objects gain cultural and aesthetic significance through careful design; and for their role in preventing access to the limited spaces of the 'war machine'. The landscapes of the active nuclear military industrial establishment are often unseen or unseeable and the traces of anti-nuclear activism lack permanence or are easily erased.[6,7] This lack of accessible materiality is just as culturally significant as the ostentatiously shiny bomb casings exhibited in Los Alamos Bradbury Museum.

Some interesting work has been undertaken to explore the material geographies of nuclear warfare in the UK. For example, geographer and artist Michael Mulvihill has spent time in residence at RAF (Royal Air Force) Fylingdales, in the UK, home of the four-minute warning, to experience the relational materiality of this place and create tactile objects, such as hand-fired missiles, while considering the assemblages that link disparate materialities and technologies, such as vinyl records, with British nuclear warfare preparedness.

POST-NUCLEAR LANDSCAPES

Geography can be used to understand the histories of landscapes, a way to explore and interpret the current significance and afterlives of nuclear places and spaces. Post-nuclear landscapes can be redeveloped, left in situ to be claimed by nature, or intentionally turned into wilderness. This has captured the geographical imagination, and there is a considerable body of literature about the materiality and environmental nature of the afterlives of nuclear places and spaces.[8] There are post-nuclear landscapes spread across five continents. They include those that have been scarred by mining and exploiting the materials needed for nuclear weapons, from the uranium mines of the Democratic Republic of the Congo in Africa, to the land surrendered for uranium mining in Nalgonda District, India, and beyond.[9] It includes places and spaces that have been used by the nuclear military industrial establishment for manufacturing nuclear weapons.[10] It also includes the places that were used for nuclear weapon testing, from French Polynesia to Kazakhstan.[11]

While Rocky Flats National Wildlife Refuge near Denver, Colorado, is today a spectacular nature reserve, its history is inextricably entwined with the US military industrial complex.[12,13] Dig beneath this place and a complex socioenvironmental history is made evident by contamination in the soil, and by the health challenges that have been experienced by the local community. This legacy began at the moment of inception of the Rocky Flats plutonium pit manufacturing plant in 1952.[14] Plutonium pits or triggers provide a catalyst for the detonation of nuclear weapons and include a beryllium starter mechanism. They are not a 'clean' item to manufacture, but rather a toxic product that creates toxic by-products. Current US legislation protects the public from harm through on-site and off-site environmental regulation. However, in the 1950s Cold War patriotism took precedence, environmental concerns were less pressing, and public awareness of the socioenvironmental effects of nuclear weapons was limited.[15] The Cold War precipitated an arms race that the Rocky Flats facility was designed to bolster by rapid production of this essential item for nuclear warfare. It is now designed for tranquillity and reflection, as part of a wider collection of Colorado Cold War heritage trails.[16]

REVIS(IT)ING THE ATOMIC BOMB

People want to explore the landscapes of nuclear warfare. Some have an interest in history, some want to memorialise the experiences of affected family members, some have worked in these places and spaces, and others want a dramatic and different holiday in an era of individualism.

From war to peace; from Hiroshima to Nagasaki; from Los Alamos to Nevada, from Maralinga to 'Christmas Island' (Kiritimati).[17] Considerable work has been undertaken to understand the dark tourism of places like Chernobyl, and the motivation and meaning of recreationally visiting post-nuclear places.[18] Tourism causes the dereliction of post-nuclear spaces to regain significance, but it also develops in them an element of theatre. They become repurposed and repositioned to both recreate and reinvent history, presenting new and adjusted narratives to militarised pasts. Woodward astutely identifies that Cold War sites are 'increasingly wrapped in interpretative frameworks which portray nuclear weaponry and war as a feature of the past'; they are avoiding discourse arising around the contemporary challenges of nuclear warfare.[19] She also notes that Cold War sites are embedded in historic narratives within broader heritage management regimes, despite nuclear weapons remaining a threat.[20]

Recently, work has been undertaken to explore existential tourism by the atomic veterans. In 2018, 60 years after the British H-bomb tests, a group of 80-year-old veterans travelled across the world to return to Kiritimati.[21] Some wanted to revisit and make sense of their past, to show their wives what it was like, and to try to understand changes to the space and landscape since their occupation. Others were concerned for the welfare of the islander community, viewing themselves as akin to those nuclear 'victims'. They attempted to retrace their pathways through a long-changed landscape, discovering relics of their time on 'Christmas Island'.

Other post-nuclear spaces provide their own unique tourism phenomena. Los Alamos is an important component of the Manhattan Project National Historical Park.[22] The canyons of Los Alamos have been part of the US attempt to re-wild and disguise the influence of the nuclear military industrial complex upon the natural landscape. Acid Canyon used to contain waste from Los Alamos Nuclear Laboratory (LANL), but is now a pretty grove of pine trees, a place for local children to hike and explore – there is even a nature centre. Local museums and mobile apps can show you the city of Los Alamos as it used to be. Tourist attractions include the Los Alamos visitor centre, LANL's Bradbury Science Museum, Los Alamos Historical Museum, and Fuller Lodge Art Museum. These places retell the story of the atomic bomb, capitalising upon the culture of secrecy that surrounds nuclear weapons to rework the public-facing history of Los Alamos. To some extent, cultural memory is constructed and transformed to preserve nuclear interests. While exploring the museum myself, I was approached by curators who provided me with more information about exhibits. Emphasis was upon the physical processes behind nuclear weapons and their necessity to end the Second World War, rather than the social and environmental science that surrounds the development, and the aftermath of the atomic bombing of Hiroshima and Nagasaki. Despite both curators during my visit being women, it felt like a very masculine space. Decorated in greys and oranges, it contained sculptures of prominent men, barbed wire for dramatic effect, interactive displays about the history and research of LANL, and hunks of bomb casing. Focus was upon the male scientists and engineers who enabled nuclear technology, the processes of defence, and a small section devoted to current research. An apt reflection of the predominantly masculine domain of war, with the role of women obscured.

This small town is peaceful, with relatively little crime, but references to war and nuclear weapons are still everywhere. Nuclear themed beers and salsa can be bought in the local bars and shops, and Atomic City Quilts provides craft and sewing apparel. Streets are thematically named: Trinity Drive, Oppenheimer Drive, Black Hole Way. This space has been transformed from a secret city into a tourist attraction, enticing thousands of visitors each year. The only thing that marks out LANL as different is the process of driving there. I must quickly stash my DSLR in the boot – all cameras are forbidden on the approach. Otherwise, the accessible areas feel much like any other academic campus, with green grass, coffee shops (I have coffee in the Otowi Café with historian Dr Alan Carr), and young people strolling about completely at ease.

Other forms of nuclear tourism occur in the form of visiting the architecture of civil defence and nuclear war preparedness. Some places have been transformed from Cold War architecture to historical tourist attraction, such as the Churchill War Rooms in London. Old nuclear bunkers that haven't been repurposed have an eerie, abandoned feel. As the history of British civil defence is slowly forgotten, we have literally attempted to bury the physical traces of our old fears of the Eastern bloc. As abandoned facilities quietly decay, they often fall into dereliction, and the materials they are constructed from present contamination risks. While it is easy to become nostalgic about the idea of somewhere safe, many of the bunkers dotted about the UK, US, and Europe during the Cold War were not designed to save the lives of ordinary citizens, just of

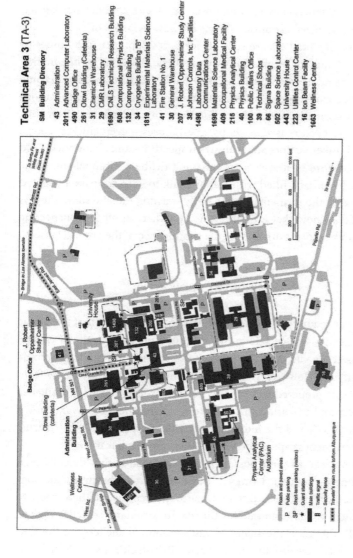

Figure 5.1 The current public geography of LANL reveals little about this place – exploration and conversation are needed to understand Los Alamos

Source: Visitor's Guide, LANL: https://legacy.lib.utexas.edu/maps/united_states/los_alamos_guide98.pdf

politicians and military personnel – the apparatus of the nation-state. Some surviving bunkers have been repurposed as emergency control centres, which house defence control centres equipped to respond to a national emergency. This seems like an apt use for local government bunkers since emergency planning originated with civil defence legislation. These bunkers' filtration systems are now switched off.

Occasionally, bunkers are sold off for redevelopment into homes or museums. A nuclear bunker in Northern Ireland was put on the market for just £575,000. Elsewhere in Europe, bunker architecture has become a part of the landscape. The German countryside is dotted with various facilities informally open for exploration. In Albania and Romania, shelters have been reused as shops, guesthouses – even chicken coops. Informal bunker visitors aren't always respectful – and abandoned Cold War architecture is at risk of trespass and deliberate vandalism. Some sites have been back-filled with soil or chalk to keep trespassers out. At Beachy Head, the former site of an early warning radar system, there is little evidence now that a bunker was ever there at all – unsuspecting tourists walk across it unaware of the significance of the ground beneath their feet.

SURVIVAL SOCIETY

Nuclear war requires meticulous preparedness, to ensure that the public accept that they will be safe in the event of a nuclear attack. Of course, no one is truly safe, and the outcome of nuclear attacks is gruesome and horrifying. However, two interesting but linked cultures grew from nuclear war preparedness: the government-level civilian resilience instruction and placation that occurred during the Cold War, known as 'civil defence'; and the individual activities of 'preppers' who distrust their government's capacity to protect them from nuclear war.

Every nuclear weapon possessor state has had a civil defence plan at one time or another. However, there are some substantial 'ideological differences' between different states' perception of their need to protect their public from nuclear threats. In command societies such as the USSR, this took the shape of vast communal shelters. In the UK, flimsy leaflets were distributed to instruct British citizens to 'Protect and Survive', despite there being no public provision for surviving a nuclear attack.[23,24]

The first civil defence plans were developed during the Second World War and were not intended to protect against nuclear threats. New measures were implemented during the early Cold War by the USSR, US and UK to reassure the public as much as to prepare for nuclear war. For governments, the big bonus of creating civil defence plans is that they could be leaked to the other side and give a false representation of the size and sophistication of their nuclear arsenal. This meant that both the USA and Russia created a sophisticated network of drills, bunkers and 'safe' spaces for the public, to give the illusion of greater proliferation than the other.[25]

You can imagine the fear that must have proliferated in the USA and the UK during the Cold War, symbolised by the B52 bombers and nuclear submarines that were ready to attack the communist bloc.[26] The US National Security Archives recently declassified the 1959 Atomic Weapons Requirements Study, providing startling insights into the trigger-happy mentality of Cold War America. The study showed that major cities, including East Berlin, were prioritised for 'systematic destruction' by nuclear attack. Hydrogen bombs, the most powerful nuclear weapons, were to be used against 'air power' targets, and to attack populations in the USSR, China and Eastern Europe, in violation of international legal norms.[27] With strategies like these in the West, everyone on Earth was made vulnerable, and every nuclear weapon possessor state needed a good civil defence plan.

However, this was not possible for many states. By the late 1950s, recessional Britain hadn't the funds to create a practical and meaningful civil defence programme, and local authorities were instructed to 'consist primarily of planning how to raise the level of preparations, should the circumstances demand it ... rather than of making physical preparations against the contingency of an imminent war'.[28] The British government decided that British civil defence would protect the state instead of society, in case of nuclear warfare. Bunkers were built for government and other 'essential' elites, whereas the public were essentially left to fend for themselves, instructed to create dens, stockpile food and paint their windows white. Public civil defence advice created a new 'citizen's architecture' of sandbags, doors removed from hinges, and fallout protection 'walls' made of suitcases stuffed with books. These are obviously not realistic or effective preparedness measures for a nuclear attack.

This created an outcry and some local authorities refused to undertake civil defence planning measures, becoming 'nuclear-free zones'.

Manchester became the first city in the world to declare itself as nuclear free in November 1980, and others soon followed suit.[29] This had significant implications for national civil defence exercises, and the refusal of 24 county councils to cooperate with exercise 'Hard Rock' in July 1982 resulted in its cancellation.[30] There was public outcry at this fiasco, and further civil defence preparedness measures were viewed with cynicism by local government and the public alike. It was evident that state coordination was inadequate and fatalistic.

This individualistic British experience was not reflected across Europe. Switzerland was an example of one of the few Western European countries that did commit to a wholesale bunker-building programme, but, consistent with the general trend towards the privatisation of civil defence, did so in a way that placed legal obligations upon developers of houses and apartments to install basements shelters, and encouraged the retrofitting of such facilities into existing buildings via tax incentives.[31] Israel's shelter policy shifted from state provision of communal shelters to the mandating of private shelter construction within new developments,[32] and of encouraging home owners to see them as home improvement, creating extra functional space in peace-time and adding value to their homes.

After the Cold War, civil defence was revamped and presented to the UK as 'emergency planning', an all-hazards approach to risk. In the UK, this is driven by the Cabinet Office rather than having its own agency, as Russia does with EMERCOM and the USA does with the Federal Emergency Management Agency. As for Cold War civil defence, UK emergency planning places an emphasis on local plans for community resilience, rather than top-down state protection of its people.

Most of the bunkers, buildings and monitoring posts that survive from the Cold War are too decrepit to resurrect. They are decaying monuments to government policies of control, deterrence and economic restraint over society's survival. In the event of a nuclear attack today, these crumbling legacy pieces would not help us. As during the height of the Cold War, most of us cannot reasonably expect to survive a nuclear attack and remain startlingly unprepared.

Preppers

The advent of the second age of nuclear warfare has left some people worried about their personal safety and wellbeing. They have recognised that state preparedness for existential threats such as nuclear warfare is minimal and have decided that they will have to try and prepare by themselves.[33] Preppers, survivalists, or doomsayers are the people who have decided to try and ensure their own survival, despite the future looking dark.[34] Prepping has emerged from the ashes of civil defence, Cold War nuclear attack preparedness and the contemporary neoliberalisation of disaster preparedness.

This is not a new phenomenon in the UK. In the early 1980s, hobbyists were guided by the short-lived magazine, *Practical Civil Defence*. In the absence of state-led shelters, they sought to build their own. One editorial remarked that while politicians procrastinated, men of 'iron will' and 'hard won engineering skills' would take on the challenge themselves.[35]

There are many ways to prepare for the apocalypse. This unique subculture has created a new disaster capitalism, buying and consuming everything from ration packs to deluxe bunkers. In places like Silicon Valley, the wealthy are creating secure spaces for themselves, hunkering down in gated communities, installing panic rooms, and excavating basements.[36,37] The rise of the 'bourgeois bunker' – high-tech, luxurious, private bunkers to protect the concerned and wealthy – demonstrates that, as ever, only the moneyed, connected and powerful are entitled to outlive nuclear catastrophe. As they try to create secure and defensible spaces to protect themselves from other survivors, younger and less wealthy preppers 'tool-up', arming themselves with bush-craft and survival skills.[38,39]

Traditional preppers with bunkers and food stocks are often white, male and seeking opportunities for defensive withdrawal into a place and state of individualistic self-protection. It has been described as a symptom of the 'anxiety of affluence', an expensive hobby of accumulation and hoarding undertaken primarily by rich libertarian Midwesterners. These activities may be undertaken under the guise of nuclear warfare preparedness, but are actually just another form of consumption for status.[40] There is a profit in supplying all the post-apocalyptic kit and technology, and ultimately the participants 'are armed to the teeth and ready to kill their neighbours'.[41] Their behaviour reflects the overarching change in resilience culture, where preparedness has become more individualised and less supported by the communal shelters of a Cold War command economy. This does not paint a very attractive picture of prepper culture, but this only speaks for the wealthier portion of this subculture.

While all preppers seem poised to leave traditional society behind and begin a new life off-grid, not all are focused on the materialities of nuclear war survival. I explored the lives and concerns of other types of prepper while in the American Midwest in 2016. Specifically, I was interested in the younger and less affluent ones. I discovered that the latest generation of preppers are less right-wing and wealthy than their predecessors and were trying to prepare for a new era of Republican politics in the wake of Trump. These people were less concerned about macho culture and the accumulation of stuff, and instead expressed deep concerns about the existential threats posed by malicious technologies, nuclear war and climate change. There has been a generational shift towards community support and skillset development, viewing resilience as something that needs strong community networks rather than being individualistic. The preppers that I met were sociable and socially aware, interested in ethical living and environmentalism.

The perspectives of this community did have an element of self-interest, with one participant saying: 'If you have a stockpile of things, then you'll become a target for those things; whereas if you have a stockpile of skills, you'll be the person that people work to keep alive … I'm not a big fan of guns in general.' Another participant, who offers free training in survival skills, told me that he did this 'to teach people how to take care of themselves and be prepared'. He described the traditional idea of survivalism as 'extreme … a lot of those people are anti-government, they're just waiting for the apocalypse or rapture. Some of them are conspiracy theorists.'

It is understandable that communities feel increasingly insecure. Prepping has become more mainstream as nuclear geopolitics becomes more volatile, and the possibility of nuclear war encroaches upon reality. Civil defence measures for nuclear war preparedness have returned, and the occurrence of false alarms of nuclear attack, mean that it is hardly surprising they are preparing for the worst. In these geopolitically interesting times, it's difficult to discern the true level of risk or the safest place to be. Even if we are prepared for the worst, our options are fewer than we might realise.

6
Strange Cartographies and War Games

> Look at your map ... This is a new kind of war. It is different from all other wars of the past, not only in its methods and weapons, but also in its geography.
>
> – Franklin Delano Roosevelt, 23 February 1942

Maps, models and war games are all part of the practice of preparing for an impossibly bleak nuclear future. They have been used over the last 70 years to justify, and to try to evade, the existential threat of nuclear warfare. The maps and models produced are interesting, because they are presented as unbiased and factual figures – but are often deliberately designed to mislead and bamboozle, to create an ideological or emotional response in the viewer or user.[1]

For instance, 'Wild Bill' Bunge's *Nuclear war atlas* is an unusually politicised series of Cold War maps that depict a desolate world after nuclear attack – complete with symbologies of disintegrating radioactive zombies stumbling through derelict wasteland. The perfect example of one person's radical ideologies being represented as an entire emotional cartography.[2] His work was important, because it provided a rebellious aesthetic for the risks that were keenly felt by many during that time.

The second age of nuclear warfare that has arisen since the end of the Cold War is characterised by multiple nuclear weapon possessor states, a wider nuclear umbrella, changing allegiances, and multilateral nuclear threats arising from several states simultaneously.[3] This has meant that maps, models and games have had to be adapted to match the current nuclear scenario. Technological advances have meant that while some maps and models provide more realistic and accurate representations of risk, more sensationalist scenarios with alarming titles quickly spread across online tabloid news venues.[4] This is one of the ways that the nuclear anxieties of the Cold War, discussed in Chapter 5, have returned.

Nuclear war computer games are currently in vogue. Every day, millions of people sit down to play at apocalypse, emerging to blink at the hazy sunlight from nuclear bunkers, before undertaking quests across desolate radioactive landscapes on their games console or PC. The legacy of our virtual annihilation fantasies is embedded in the war game strategies of the past. Game theory was an important component of negotiation during the Cold War, providing an early warning of the possibility of mutually assured destruction (MAD), if nuclear war escalated between America and the USSR. Simulations are now used with varying degrees of complexity and accuracy to represent the outcomes of nuclear war by government agencies, academics and amateur map makers. Both historic and contemporary geographers have a role in creating and imbuing meaning in geotechnologies – and understanding the links between simulated nuclear war and 'apocotainment'.[5]

GEOGRAPHERS AT WAR

Nuclear war would have been considerably more difficult without geographers. Lacoste declared that 'Geography serves, first and foremost, to wage war.'[6] Early works by geographers of nuclear issues contained unbridled optimism about the potential for nuclear technology to reinvigorate a European continent that had been ravaged by the Second World War.[7] The discipline originally had a military heritage and a colonial stance, and much research in geography is still funded by the military. This develops and maintains skills that are important to warfare such as Geographical Information Systems (GIS), spatiotemporal and risk modelling. However, these skills are equally useful to pacifism and disarmament, and some geographers have recognised their capacity to create peace, pushing back against geography's warlike origins.

During the early to mid-Cold War, nuclear war technologies flourished internationally, and unwitting geographical harm was done to humans and environment. This included the work of 'Project Ploughshare' between 1957 and 1973, which focused on 'geographical engineering', modifying the physical geography of landscapes through the 'careful' deployment of nuclear weapons to scar the land.[8] However, as the public and academic community became aware of the harm caused by nuclear weapon tests, then a more critical and subversive approach to nuclear weapons began to emerge in geography.

Pacifist and critical geographers existed before nuclear weapons. In 1941, during the Second World War, German émigré sociologist and early 'radical geographer' Hans Speier wrote an essay in *Social Research* titled, 'Magic geography'. In this essay he argues:

> Maps are not confined to the representation of a given state-of-affairs. They can be drawn to symbolise changes, or as blueprints of the future. They may make certain traits and properties of the world they depict more intelligible – or may distort or deny them.[9]

Speier recognised the capacity for maps to be used for harm, long before nuclear weapons became part of the geopolitical agenda.

Geographers began to critique the social impacts of nuclear deterrence during the late Cold War, and this helped to critically disrupt the discipline's links to war and empire by creating new and radical ways to explore nuclear weapons. Both Wisner and Cutter's critical descriptions of nuclear warfare in the 1980s provided important insights into the need for pacifist geographers' voices to be heard, for the geographies of violence and injustice to feature prominently on research agendas, and for the social responsibilities of the geographer to be recognised.[10,11] Cutter's work is notable; she pioneered ways to discuss nuclear weapons through the context of peace, declaring that 'issues of nuclear war and deterrence are inherently geographical',[12] before becoming the leading expert of her time on existential threats posed by technological hazards.

The work of these geographers helped to launch a flurry of research into the impacts of nuclear warfare that is still relevant to this day. The scope of critical work from the late Cold War includes studies of the potential human impacts of hypothetical nuclear attack on UK cities, the risk of nuclear power plant sabotage, and the long-term consequences of nuclear waste sites.[13,14,15] These spatial studies of nuclear attack by Openshaw and his collaborators redeployed maps and models to critique the human consequences of nuclear warfare in the UK.

Bill Bunge's Reagan-era *Nuclear war atlas* has already been mentioned, as it explored the cultural hysteria of the Cold War through dystopian cartography.[16] His atlas visually explored the implausibility of human survival after nuclear war by using a striking red and black colour scheme and startling symbologies. His eye-catching choice of colour scheme was not new, it was instead repurposed from the blast zone maps that were created after Hiroshima and Nagasaki. He spatially represented studies from the 1970s to generate social change, rather than to present a strictly accurate perspective on nuclear risk. The absurdity of his maps – with the blind, sick and insane all dying amid 'wiggly lines of radiation',[17] visually cemented the irrationality of nuclear warfare.

It is still challenging for geographers to platform their pacifism, and there is still a lack of geographical influence on nuclear policy. A trend of vaguely mapping and modelling nuclear warfare has re-emerged in recent years – although the motivations behind this work are now darker, tinged with the need for shock value and the exhibitionism of the social media age. Maps such as Wellerstein's playful online Google Maps mash-up NukeMap have gained publicity by presenting a new public hybrid of technology and game. Using Nuke Map, it is possible to hypothetically and unthinkingly destroy cities in nuclear blasts, creating nuclear war 'what if?' fantasies.

CARTOGRAPHIES AND TECHNOLOGIES

History is dominated by images of nuclear physicists pioneering the first atomic bomb, but geographers played a key role in nuclear warfare by developing geotechnologies. Current nuclear weapon systems are reliant upon geotechnologies that were developed during the Cold War, from GIS, Global Positioning Systems (GPS) and Remote Sensing; to other spatial modelling approaches for the control, monitoring and deployment of nuclear weapons.

Geotechnologies have also been used for nuclear warfare preparedness. Over time and through the use of maps and figures alarmist campaigns such as the UK government's 'Protect and Survive' civil defence leaflets, this has created an image problem. Imprecise solid concentric rings around a target location are the most common misrepresentation of the consequences of nuclear war, as they create an aesthetically pleasing output for nuclear apocalypse. As technologies have improved, the representation of nuclear warfare has not. The commonality that binds maps of nuclear war together is the assured certainty of their cheerfully coloured concentric outputs, in promising doomsday or safety with authority, if not accuracy.

It is the responsibility of geographers to ensure that cartography and other geotechnologies are used to support our understanding of nuclear warfare, rather than merely representing the ideologies and anxieties of those who commission and create maps.[18] There is a need for greater engagement with the public to support everyday understanding of maps. For example, has a tabloid newspaper consulted an expert for its blast radii map, or is this an example of misinformation, designed to boost readership and perpetuate new mythologies of risk?[19]

These maps are presented in contrast to the cartographies that followed Hiroshima and Nagasaki, where the blast zones of destroyed cities became simplified and dehumanised blocks of pink and red, black and white.[20] Perhaps this simple colour scheme made the atrocities of nuclear warfare easier to process. Further research is still needed to examine and verify the substantial yet patchy archive of nuclear warfare cartography from the Cold War. Each map is cluttered with both symbology and symbolism, imbued with meaning like an old painting. We need cartographically literate people to explore these unduly respected artefacts, and to ensure that the visual trickery and biases that may be present in contemporary infographics and maps of 'nuclear warfare' are appreciated. Only then can the covert cartographies of nuclear warfare be truly understood.

However, there is substantially more to the geography and geotechnologies of nuclear warfare than just maps. In fact, the advent of rocketry was basically Cold War geopolitics under another name.[21] The development of new spatial analysis geotechnologies fixed a host of challenges, from the 'bomber gap' of the 1950s to the 'missile gap' of the 1960s. Many that are now commonplace in the civil domain were first developed in military secrecy and have since been repurposed.[22,23] From applied spatial modelling, to the analysis of satellite imagery and the creation of geodatabases to store spatial data, all gained significant technological advances due to nuclear warfare.

Geotechnologies also had a surprising role in preventing the outbreak of nuclear war by improving monitoring and surveillance. The Cold War was kept chilled by surveillance and spying, including imagery intelligence collected by photogrammetry and satellite remote sensing that gave international reassurance. This approach gave covert insights into the number, type and location of conventional arms, making arms control a reality and reducing the total threat of a global nuclear war. However, while insights were provided into enemy territory, there were still mysterious zones that remained off-limits to geotechnologies, and their presence dwelled in the political imagination.

New geotechnologies were developed during the Cold War to try to close these gaps and reveal ground truths and home truths, including thermal imaging, laser range finding, digital mapping and the internet. The technologies of nuclear warfare also created a vertically militarised geography, where space was weaponised for surveillance. A state's capacity to send a rocket into space also paralleled its capacity to deploy nuclear warheads across the globe. One sensible outcome of the space race was the 1967 United Nations (UN) Outer Space Treaty, which banned nuclear weapons in space, on the moon or on other celestial bodies, and contained a directive to use outer space for peaceful purposes.

Over the last 20 years, new military technologies have emerged in geographically connected domains, including advances in human factor modelling, agent-based modelling, 3D and 4D mapping, unmanned aerial vehicles (UAVs) and drones, and big data for understanding risk. There are digital command-and-control systems, anti-satellite weapons, geospatial simulations and precision-guided munitions. This automated nuclear warfare needs geotechnologies and critical geographers, now more than ever before. We are in a post-cyber nuclear age, where it is now possible to simulate the testing and deployment of nuclear weapons. Both nuclear weapons possessor states and nuclear weapons desiring states can exploit monitoring and surveillance geotechnologies, including GIS and seismography. Surveillance data can be manipulated to 'prove' or disprove their possession of nuclear weapons. Systems can be hacked.

IT'S JUST A GAME

From games theory to apocalyptic computer games, ludology, or the study of games, has played a major role in understanding the culture that surrounds both real nuclear warfare scenarios and the apocalyptic imaginary. Play, in the form of scenario enactment and desktop exercises, has often been used by civil defence and governance to understand the potential outcomes of nuclear warfare.[24] From there, the idea of simulated nuclear war, or playing at nuclear war, has slipped seamlessly into the recreational gaming arena.

There is a commonality between all forms of nuclear war game, from contemporary dystopian gamer scenarios, to the work undertaken by the RAND Corporation that explored the outcomes of different attack scenarios in the 1960s. All games offer the player the opportunity to reshape possibilities through geography and geopolitics, through policy or strategy adjustments at government level, to imagining what it would be like to survive nuclear war in a computer game.[25] Computer games are an important element of this nuclear imaginary, as they offer a shift towards a *culture of simulation*, where digital technologies make it possible to create, interrogate and destroy hypothetical game-worlds.[26]

The phenomenon of playing at nuclear war is interesting, as it provides insights into what is considered appropriate to move from the military to leisure domain. Like paintballing or assault courses, apocalyptic computer games give militarism a shiny veneer of acceptability and invite complicity.[27] They allow nuclear warfare to become friendly and hospitable. They make the public familiar with military programmes, missions and technologies. To some extent, the familiarity of games like *Fallout* justifies and legitimises nuclear warfare, unshackling the nuclear taboo. The nuclear defence sector is no longer greased by secrecy, but by games.

Gaming was born from warfare. Gaming technology originally emerged from military activities, and the two sectors feed into each other.[28] The RAND Corporation was a key player in the creation of nuclear war game theory. Their work included the development of MAD theory, which was applied to nuclear war scenarios by futurist and Cold War strategist Herman Kahn with grim results in 1962. This, and geotechnologies such as topographic modelling and environment generation, fed into the production of games for leisure rather than warfare. Some of the earliest games, such as *Missile Command*, presented a realistic narrative of nuclear anxiety.

National Command Level (NCL) decision-making tools were also developed from game theory to explore the outcome of different superpower interactions. Their purpose is to aid or stand in for human players in political-military war games, which are conducted for training, or to explore new strategic concepts and consider alternative views of deterrence, escalation control and war termination. A new programme of NCL models is currently being developed by RAND Strategy Assessment Centre using game-structured simulation systems to improve understanding of plausible scenarios. The scenarios may 'begin with crisis and extend through general nuclear war, or ... may begin with crisis and terminate without cataclysmic warfare'.[29] So, where does nuclear warfare preparedness diverge into leisurely ludology, and what does this mean for a generation of gamers who have grown up on both potential and virtual nuclear war?

PLAYING APOCALYPSE

There is a marked overlap between simulations and games, but also some key differences. The chief differences are accuracy, intent and purpose.

Simulations are designed to inform, whereas games are designed to entertain. The nature of computer games can tell a great deal about the nature of nuclear culture, and many of these games seek to represent and celebrate warfare of many types, from *Call of Duty* to *Gears of War*.

The gaming industry has grown rapidly in recent years, becoming one of the largest and most influential forms of popular entertainment. Postmodern society has never been in more desperate need of entertainment and escapism than it is now, and gaming offers a way to forget the insufferable dreariness of neoliberal life, to engage in something more winnable than late-era capitalism, even if that means 'surviving' in a desolate post-apocalyptic radioactive wasteland.

These games do not represent a solution to a broken reality, but they provide just one example of many possible coping mechanisms that people have become unwittingly and compulsively dependent upon for their psychological survival.[30] They offer an opportunity to play through the anxieties that accompany a time of great geopolitical instability.[31] They provide alternative realities, often those that denounce consumerist materialism. Ironically, they are often sold through corporate ideology. They harness the ideologies of subversion as a marketing power. Post-apocalyptic gaming is notably very white, very male and very middle class, as storylines focus upon individualistic self-protection and self-resilience. Like real-life survivalists, gamers are served by commercial suppliers of survival kit, such as *Fallout*'s catchily named 'Pipboy' monitors and 'Rad-X' ionising radiation exposure treatments. Ultimately, gamers live within a violently libertarian virtual society, while 'armed to the teeth and ready to kill their neighbours'.[32]

Apocalyptic games generally have macho titles that evoke desolation and hopelessness, and that repurpose ideas of radioactivity and risk. Titles include *The Wastes*, *DEFCON: Everybody Dies*, *Fallout* and *S.T.A.L.K.E.R.* It would take an entire book to explore the geography of every nuclear war game. However, *Fallout* and *S.T.A.L.K.E.R.* are both popular and populist choices, and they offer interesting insights into the geography of post-apocalyptic gaming, where the atom has become the dominant technology.

The *Fallout* series offers a post-apocalyptic and action-filled world where decidedly human endeavours play out. Set in the decades and centuries following the aftermath of a nuclear war, this game envisages life within the factional and militaristic societies in a ruined and divided USA. *Fallout* may be set in a contaminated radioactive wasteland, but it is still a relatable and nostalgic dystopia. It is an interactive model of a failed society, complete with retro-futuristic features reminiscent of the *Plutopian*[33] idyll of 1950s USA. It offers a critique of picket-fenced capitalism that occurred across the US.[34] The geography of *Fallout* consists of vaults, ruins and wastelands – despite the post-apocalyptic environment being dubbed a wasteland, there is a relic of human presence in every frame.

The *Fallout* series takes the player on a tour of dereliction across California, Chicago, Texas, Washington DC, Las Vegas, Utah and Boston. The Washington DC scenario of *Fallout III* includes a nuclear war game simulation, a meta addition, that plays on the evolution and significance of the war game and simulation to playing apocalypse.

The *Fallout* world is not completely irradiated but contains hotspots of severe contamination at missile impact sites. The world is viewed through an interface that includes technologies that detect radiation, including dosimeters and a Geiger counter. Dose is described in Rads rather than the more commonly used Sieverts or Greys. The exposure likelihood matrix has evolved during the series but, unsurprisingly, does not reflect the scientific realities of radiation exposure. However, the accumulation of dosage does lead to 'radiation poisoning' akin to acute radiation syndrome (ARS). There is an embodiment of the experience of ARS, as the player becomes progressively more unwell with increased exposure. An element of both corporate pharma and medical sci-fi is included here, as drugs including 'RadAway' and 'Rad-X' can be purchased in the game to treat and recover from the effects of excess radiation exposure.

The protective vaults in *Fallout* reflect the multitude of fallout shelters and bunkers of the American civil defence movement,[35] with most of them safely containing their human subjects until the radioactive dust had settled. However, this is where the bunker realism ends. *Fallout*'s approach to civil defence has more in common with the British legacy of civil defence, designed to preserve the state rather than society. Each vault has a 1,000-person capacity, but there are just 122 self-sufficient vaults for a population that reached 400 million by 2077.[36] Access to the vaults was gained by payment or winning a lottery, meaning that only those who could afford it were able to survive.

The subterranean geography of *Fallout* is extensive, with some vault-dwellers' lives occurring entirely underground.[37] However, this life is demographically augmented by Vault-Tec, an 'evil' corporation who design both bunkers and social experiments. Some bunkers have skewed ratios of male to female occupants, and other torturous and unethical 'psychological experiments' are in train, such as arming residents, remaining closed for 200 years to prolong isolation, and selecting people with radically diverse ideologies to live together permanently.

Vault-dwellers are not the only surviving humans. *Fallout* has subalterns who are trapped within a caste system of social stratification. There are four castes, in a system that plays heavily on the perceived stigma and health effects that are associated with ionising radiation exposure: a privileged elite minority of vault-dwellers who have not been exposed to ionising radiation, those who survived the blast without being in a vault; those who have been exposed to ionising radiation and become 'ghouls', who are capable but treated as less-than-human and therefore wary of other people; and 'feral ghouls' at the bottom of the hierarchy, who become demented over time from ionising radiation exposure, losing their ability to reason, and becoming incoherent and aggressive.

Feral ghouls are treated like animals by humans who are horrified by them, whereas regular ghouls pity them. Ghouls are normal and can chat with the player, they are disfigured and have croaky voices, but are fully cognisant humans. Ghouls have jobs, they farm and hunt, but there is an apartheid between 'ghoul' people and other humans, despite the superficiality of visible disability of 'ghouls'. *Fallout* is played from the first-person perspective of a previous vault-dweller and involves conflict between the different pockets of society and the vault-dwellers. Therefore, the game player is placed in the positionality of the surviving 'middle-class' community, in the context of the game world.

Speaking to players of the *Fallout* series confirmed that 'radiation sickness is a strong and running theme, relevant to pretty much everything you eat, drink or swim through. A constant reminder that you're in a pretty crappy and dangerous world.' Perhaps there is an element of relief to apocalyptic gameplay, in contemplating scenarios and geographies that are worse than late-era capitalism. Players feel that 'the franchise has an underlying message of hope despite the initial, apparent horror of what we've managed to do to the world and the subsequent need to hide away in relative safety until it was deemed safe to come out'.[38]

Ebb, a woman who is a *Fallout* player, said:

One of the moments of 'light' throughout this are the radio stations. The idea is that some of the music, in the form of records, survived.

There are a few existing / surviving comms towers and as such, you can be wandering a dangerous terrain, shooting at a giant RadScorpion with a gun you've put together from various salvaged bits, loaded with salvaged bullets, listening to Cole Porter's 'Anything goes' ... There are some very tongue-in-cheek choices by the game developers thrown in ... 'I don't want to set the world on fire' by The Inkspots, 'Orange coloured sky' by Nat King Cole, 'The end of the world' by Skeeter Davis.[39]

She went on to talk from a positive perspective, saying that:

It keeps you mindful of the fact that humanity has been, and remains, on the brink of oblivion regardless of whether that's post-holocaust or because of global warming. But it does it in an engaging way. As for what makes me think 'Ah, let's have a bit of escapism ... I know, nuclear holocaust time!', it tends to pull me in when I need to laser-blast something or want to fill myself with a false and vain hope that somehow we'd, long term, be ok in the end. I find it reassuring and cheering.

The tragic irony of the *Fallout* series is, of course, that it tells a cautionary tale of a fatally complacent society that the commodification, globalisation and consumerism of gaming is helping to create.[40]

Nuclear simulation and play provides a way to return from the brink through a simple restart. The real-life consequences of nuclear deterrence include no caveats for 'replay', despite civil defence emergency planners' aspirations of post-strike response and recovery. Nuclear apocotainment such as *Fallout* seeks to 'normalise' a dystopia that is not merely a desolate wasteland, but also features masculinity, privilege in the form of the entitled classes of the vault occupiers above all others, and neo-colonialism. Games such as *Fallout* capitalise and merchandise racism, poverty porn and, above all, building a nation that is lacking in empathy and welcoming the certainty of doomsday, in anticipation of being able to kill roaming marauders in real life, happily and without consequence.

The depiction of virtual ruins of the present in games such as *Fallout* simultaneously reproduces the empty novelty of the commodity- and progress-oriented civilisation, while offering a vision of failed progress through counter-spectacle. It reveals the illusion of progress as fallacious and, in contrast to commodity capitalism, counter to the dream of eternal progress.[41] Games like *Fallout* fuse representational, corporeal and affective dynamics to simulate geopolitical order, but also create experimental and disruptive capacities.[42]

This idea of survivalist notions spills out into mainstream culture. It is arguable that these films, games, maps and models make us more likely to envisage a dystopian nuclear future and to begin to prepare for the end. From *Dr Strangelove* to *Mad Max* and *Planet of the Apes*, nuclear apocalypse is a movie genre in itself. One would hope that geopolitics could provide some hope and truths to counter the irregular nature of maps, models and games and their presentation of apocalyptic scenarios.

7
Spaces of Irregularity

> The nuclear arms race is like two sworn enemies standing waist deep in gasoline, one with three matches, the other with five.
>
> – Carl Sagan

Nuclear norms are the expected behaviours of arms control. The most significant norms associated with nuclear weapons are those of non-proliferation, non-use and deterrence.[1] Both vertical and horizontal proliferation are delegitimised by non-proliferation treaties, where horizontal proliferation is the spread of weapons from state to state, whereas vertical proliferation is an increase in the number of weapons in a state's own domestic arsenal. The non-proliferation norm attempts to prevent nuclear weapons from spreading to any other actors; the non-use norm sets an internalised moral restraint on detonating a nuclear weapon; and the deterrence norm provides a prudential rationale for not using nuclear weapons against an opponent who also has a nuclear arsenal.

Nuclear weapons used to be an irregular mode of warfare. It is not conventional, and not designed or intended for use in conflict, unless exceptional circumstances arise. However, the notion of the irregularity of nuclear warfare is being stretched by changing social norms that relate to nuclear weapon possession and deployment. The nuclear taboo is becoming gradually eroded, creating permanent changes to geopolitical culture.[2] Many of these changes have come about because new nations are joining the nuclear game and changing the previous status quo.

The institution of the nuclear weapon possessor state has been fixed into international geopolitics as a permanent security doctrine. They deploy their perspectives on the Non-Proliferation Treaty (NPT) as a legitimising and normative framework for their own possession and on their own terms, as a management tool to enforce non-proliferation on other states.[3] They have a diverse arsenal of nuclear weapons to ensure compliance.[4] This includes tactical nuclear weapons for direct use on the battlefield against enemy forces, and powerful strategic nuclear weapons, which are designed to create maximum damage, destroying entire cities and industries, and killing hundreds of thousands of people.[5,6] There are intercontinental ballistic missile (ICBM) and submarine systems across the world, waiting for that unlikely moment when the need for deployment arises. On paper, the policing and control of nuclear weapon ownership, or arms control, should be relatively easy for dominant nuclear weapon possessor states to attain.

Nuclear warfare has been normalised. Nuclearism, the belief of psychological, political and military dependence on nuclear weapons, and the embrace of weapons for the solution to a wide variety of human dilemmas, most ironically, that of 'security' is normal.[7] This ideology is based on a series of naïve assumptions – that the use of nuclear weapons can be managed, that their effects can be controlled, and that protection and recovery in a nuclear war are meaningful ideas.[8] Like any other form of extremism when it becomes mainstream, it makes nuclear weapons appear as inevitable and acceptable features of international security.

In the history of mankind, nuclear weapons are not a normal way to resolve conflict and will likely be viewed as unacceptable by future generations, similarly to the atrocious chemical weapons that preceded them.[9] Geopolitics and rhetoric obscure the absence of humanitarian considerations, the fact that few can make decisions about nuclear weapons, and the unequal nature of nuclear weapons ownership internationally. While existing treaties have contributed to humanitarian progress, their shortcomings and flaws mean that a new type of multilateral treaty is required.

The geopolitical dominance of nuclear weapon possessor states has recently been challenged by the United Nations (UN) Treaty on the Prohibition of Nuclear Weapons (Ban Treaty), which reconsiders the legitimacy of nuclear weapons and identifies that they are incompatible with humanitarian law.[10] This gives non-nuclear states more autonomy and agency to decide if nuclear weapons are what the world needs.

States are expected not to rapidly increase their stockpiles and only to use their nuclear weapons should a worst-case scenario occur. However, international engagement with these norms is spatially inconsistent, with enforcers such as USA and UK, and violators such as North Korea. Also, the social influence of 'legitimate' nuclear weapon possessor states is not always positive. This can affect the way other states conceptualise the value of nuclear weapons, and can even create a backlash that increases likelihood of proliferation.[11]

In 1960, Herman Khan of the RAND Corporation arrived upon the idea of mutually assured destruction (MAD) as a reason to inhibit further weapon development.[12] He declared that nuclear defence technologies had grown to such enormous proportions that it would now be possible to obliterate humankind if multilateral nuclear warfare was ever undertaken.[13] Whereas the initial era of nuclear arms development created a frenzied, normalised, and even valorised race towards weapons development and deployment, MAD subsequently stigmatised and reversed those goals, firmly establishing the nuclear taboo.[14] Taboos prohibit behaviours that are not appropriate within the moral or ethical framework of an individual community – scenarios that are so dangerous or perverse that they are almost unspeakable.

Even President Ronald Reagan recognised the stigma of nuclear weapons during the Cold War, saying that 'A nuclear war cannot be won and must never be fought. The only value in our two nations possessing nuclear weapons is to make sure they will never be used.'[15] His thoughts were echoed by UK Prime Minister Margaret Thatcher and Russian President Mikhail Gorbachev during the 1980s.

Nuclear taboos were described by Nina Tannenwald, who explored the way that 'a normative prohibition on nuclear use has developed in the global system, which, although not (yet) a fully robust norm, has stigmatised nuclear weapons as unacceptable weapons of mass destruction'. She argued that the taboo has become so widely recognised that the use of nuclear weapons – whether for tactical or strategic purposes – has become 'practically unthinkable'.[16] Her work highlights the exceptional nature of nuclear weapons, and how it manifests internationally in widely held inhibitions surrounding their use, and widespread popular revulsion. These geopolitical and social norms evolved during and after the Cold War to keep the world safe from destruction and encourage denuclearisation. However, they are currently being eroded by a new era of impulsive and bombastic leaders, including Putin, Trump and Kim Jong-un.

The resurgence of more authoritarian and militarised governance and multilateral proliferation means that nuclear warfare has resurfaced as a transnational geopolitical issue. New nuclear warfare occupies different structures, spaces and processes, and has a shifting spatiality and temporality. There is a scale, from limited to full-scale nuclear war, and each conceivable point along this continuum has different geographical and geopolitical implications. There is also a network, a virtual geography of social networks and connections that globalise our nuclear anxieties through channels of misinformation and strange truths, and that has begun to influence our geopolitics, for better and for worse.

TREATIES AND POLICIES

The Global North has created a culture of legitimacy around possession of nuclear weapons, justifying their ownership on the grounds of global security. They portray themselves as the rightful custodians of nuclear warfare, adapting forms of intervention to the changing social and global context of the age of risk.[17] This legitimacy has been constructed through treaties that are designed to make the Global North appear more risk averse and 'less willing to engage themselves in wholesale warfare than almost any states in history'.[18]

Nuclear weapon possessor states preserve their claims to power by blurring and eroding the boundary between peace and war. This helps these states to enforce and monitor treaties that protect their own interests. The international community is expected to acknowledge the USA, Russia, the UK, France and China as 'nuclear states' and all others as 'non-nuclear states'. However, this is no longer rational in an age of postmodernity.[19] This divisive approach means that while some states have been permitted to develop nuclear weapons, others, such as India, are denied despite proving to be good custodians.[20,21] A list of nuclear treaties, their creators and their purpose are included here (Table 7.1) to give perspective on both geopolitical progress, but also the perpetuation of US, Russian and UK dominance.

Table 7.1 shows the nuclear treaties that have been predominantly organised by the Global North, apart from those that create nuclear-free zones and the recent disarmament treaty, which are in bold. The Global North countries constitute 'the main actors of postmodern warfare in a globalized world'.[22] They supposedly no longer wage war; instead they create order through policing, fighting terrorism, humanitarian intervention and state-building. The persistence of nuclear weapons is a symptom of this negative peace that the Global North creates, by manipulating the war/law/space nexus to further their own interests.[23]

Nuclear weapon treaties are divided into several categories. First are those that support disarmament, non-proliferation, weapons limitation and cooperation between both nuclear and non-nuclear weapon possessor states. Second, weapon limitation treaties occur between the

Table 7.1 Nuclear defence treaties: past and present (nuclear-free treaties highlighted in bold)

Date	Title	Type
7 July 2017	**Treaty on the Prohibition of Nuclear Weapons (Ban Treaty)** 59 UN signatories. Not yet ratified.	Disarmament
29 July 1957	Statute of the International Atomic Energy Agency	Non-proliferation
23 June 1961	Antarctic Treaty	
10 Oct. 1967	Outer Space Treaty	
26 Apr. 1968	**Treaty of Tlatelolco (Treaty for the Prohibition of Nuclear Weapons in Latin America and the Caribbean)**	
5 Mar. 1970	Non-Proliferation Treaty (NPT)	
18 May 1972	Seabed Arms Control Treaty	
26 Oct. 1979	Convention on the Physical Protection of Nuclear Material	
28 Oct. 1986	**South Atlantic Peace and Cooperation Zone**	
11 Dec. 1986	**Treaty of Raratonga (South Pacific Nuclear Free Zone Treaty)**	
15 Mar. 1991	Treaty on the Final Settlement with Respect to Germany	
28 Mar. 1997	**Southeast Asian Nuclear Weapon Free Zone Treaty**	
18 July 2005	123 Agreement (India–United States Civil Nuclear Agreement)	
21 Mar. 2009	**Treaty of Semipalatinsk (Central Asian Nuclear Weapon Free)**	
15 July 2009	**African Nuclear Weapon Free Zone Treaty**	
20 Dec. 1961	McCloy–Zorin Accords	Weapons limitation
10 Oct. 1963	Partial Nuclear Test Ban Treaty	
26 May 1972	SALT I (Strategic Arms Limitation Talks)	
1972–1979	SALT II	
1972–2002	Anti-Ballistic Missile Treaty	
1 June 1988	Intermediate-Range Nuclear Forces Treaty (INF)	
11 Dec. 1990	Threshold Test Ban Treaty	
1993	START I (Strategic Arms Reduction Treaty)	
5 Dec. 1994	Fissile Material Cut-off Treaty (not completed)	
9 Oct. 1996	Comprehensive Nuclear Test Ban Treaty (not in force)	
1 June 2003	Strategic Offensive Reductions Treaty (SORT)	
11 Feb. 2011	NewSTART	
19 Aug. 1943	Quebec Agreement (with Canada)	Cooperation
4 Aug. 1958	US–UK Mutual Defence Agreement	
21 Dec. 1962	Nassau Agreement	
6 Apr. 1963	Polaris Sales Agreement	

USSR and US, and later between Russia and the US. Third, cooperation treaties are predominantly between transatlantic allies, the UK and the US. While many treaties aim to prevent other states from joining the nuclear club, there is sometimes a beneficial impact, as they can embed the positive nuclear norms of non-proliferation and non-use. It is also no coincidence that weapon limitation treaties arose in the early 1960s, in part due to the idea of mutually assured destruction.[24]

The leaders of nuclear weapon possessor states have not always espoused deterrence, notable mentions being US Presidents John F. Kennedy and Barack Obama. Obama was awarded a Nobel Peace Prize in 2009 for his promotion of non-proliferation, before he became president. He committed to 'reduce the role of nuclear weapons in our national security strategy and urge others to do the same'.[25] He declared 'clearly and with conviction America's commitment to seek the peace and security of a world without nuclear weapons'. President Obama made a historic visit to Hiroshima to lay a wreath and pay tribute to the people of Hiroshima in 2016. He wrote in the visitors' book: 'We have known the agony of war. Let us now find the courage, together, to spread peace, and pursue a world without nuclear weapons.'

Obama recognised the risks of nuclear warfare and took steps to reduce proliferation and the number of ready nuclear weapons when he signed the New Strategic Arms Reduction Treaty (NewSTART) in April 2010 with Russian President Dmitri Medvedev. This treaty committed the two nations with the largest nuclear stockpiles to a decade of more rigorous strategic arms control through limitation of strategic deployment and mutual warhead inspection opportunities. It replaced the late Cold War SALT pacts and the 1991 START I pact, and limits both countries to no more than 1,550 deployed warheads and bombs. This was a 30 per cent reduction from previous bilateral agreements.[26] However, the Nobel Committee have argued that he did not make adequate progress during his time as president, with the Nobel secretary regretting Obama's Peace Prize.[27] Regardless, his attempts at denuclearisation are undermined by his successor as president.[28] Trump described Obama's NewSTART Treaty as a 'bad deal' for the US, which has raised concerns about the future of this treaty and others.[29] Treaties are essential for arms control, and there is hope that a future success by the Ban Treaty will mean that other treaties which endorse the nuclear status quo will become void. Perhaps then the maintenance and proliferation of nuclear weapons, the sharing of nuclear defence expertise and technologies, and the exploitation of the oceans and skies of non-nuclear weapon possessor states will end.

NEAR MISSES

A Ban Treaty would prevent the considerable geopolitical turmoil that arises when something goes wrong. There have been far more near-misses than most of us would imagine, and many false alarms. Arguably, even one near-miss is too many for anyone to consider nuclear weapons to be a safe and sensible option. Historic examples of near-misses include issues with nuclear bombers and ballistic missiles, problems with ambiguous data and defective early warning sensors, misinterpretation of natural phenomena, human error, technical problems with early warning and launch systems, and staff ignoring procedures or not having received proper training.[30]

The first Cold War near-miss was during the Suez Crisis. On 5 November 1956, North American Aerospace Defense Command (NORAD) received a number of reports of Soviet aircraft over Turkey and Syria, and the movement of the Soviet fleet. The US believed that these manoeuvres signalled the launch of a Soviet offensive against France and the UK in Egypt and prepared to respond with a nuclear strike. All of these reports were either exaggerated or erroneous, and the unidentified aircraft over Turkey turned out to be a flock of swans.[31] This was followed four years later by a false alarm in Greenland on 5 October 1960. US early warning radar equipment in Thule misinterpreted a beautiful Norwegian moonrise as a full-scale nuclear attack.[32]

Thankfully, the Soviet leader Nikita Khrushchev was visiting New York, encouraging commanders to quickly realise their mistake.[33] On 24 January 1961 two H-bombs were accidentally dropped on Goldsboro, North Carolina, by a bomber with a broken wing.[34] One of the bombs broke on impact as its parachute failed. The other landed unharmed, with five of six safety device failures. Later that year, staff at the Strategic Air Command Headquarters lost contact with both ballistic missile early warning system sites and with NORAD, precipitating yet another apocalyptic nuclear panic.[35] These scenarios were not unusual, fuelling a state of global anxiety.

The most concerning near-miss scenario was the Cuban Missile Crisis, which occurred between the USSR and US from 16 to 28 October 1962.[36] This 13-day confrontation concerned Soviet objections to US Jupiter ballistic missile deployments in Italy and Turkey. In response, the USSR agreed to deploy their own missiles to protect Cuba against US invasion. The US would not permit Cuba to host the missiles and tensions rose, with both US President Kennedy and USSR leader Khrushchev being pressured to strike first. Only a last-minute agreement between them prevented the outbreak of war. Kennedy agreed to relocate US missiles in Turkey in exchange for Khrushchev removing all strategic missiles from Cuba. Unbeknownst to Kennedy, USSR tactical nuclear warheads remained in Cuba until much later.

The catalogue of nuclear errors that occurred during this crisis seems incredible. On 24 October, a Soviet satellite exploded in orbit, leading the US to believe that the USSR was launching a massive ICBM attack.[37] On 25 October, a faulty sabotage alarm system at Duluth Sector Direction Center was activated by a guard, who shot at a bear that was climbing the facility fence. Instead of signalling intrusion, the alarm ordered the take-off of nuclear-armed planes. Fortunately, communications were established that cleared up the mistake before take-off.[38] The crisis deepened. On 26 October an unannounced Titan-II ICBM was launched from Florida into the South Pacific, but no one alerted the Moorestown radar site in New Jersey, causing great concern.[39]

That evening, a US U2 spy plane accidentally flew within Soviet airspace and then ran out of fuel, only to be escorted back to the USA by Soviet MIG F-102A fighters that were loaded with nuclear missiles.[40] A US U2 spy plane had been 'accidentally' shot down over Cuba by 27 October, almost escalating conflict. On 28 October, US Navy destroyers cornered Soviet submarine B-59 near Cuba, in international waters. The Soviet submarine was struggling, with low batteries, ineffective air conditioning and rising air temperatures. The B-59 crew members had no contact with Moscow, but they did have a nuclear torpedo – Captain Savitski and Torpedo Officer Grigorievich ordered that it be launched. One Soviet sailor, Officer Arkhipov, prevented the decision to launch from being authorised. He may have saved the world.[41]

The nuclear errors did not end here. NORAD received news from Moorestown again at 9.00 a.m. that a strike was expected to hit Tampa at 9.02 a.m. on 28 October.[42] A test simulation was playing out at the same time as a satellite appeared on the horizon, but Moorestown radar site had not been informed of either instance, again creating panic.[43] While there should have been overlapping radars to confirm the appearance of missiles, the additional radars were not operating at the time.[44] The final fiasco of the Cuban missile crisis occurred the same day, when the newly appointed Laredo warning site also misidentified a satellite and informed NORAD that nuclear missiles were on their way.[45]

Few lessons were learnt from the Cuban missile crisis. On 9 November 1965, a widespread power outage in the USA created the illusion of a nuclear attack.[46] On 17 January 1966, a US B-52 bomber carrying four H-bombs collided in mid-air while refuelling over Spain.[47] Conventional explosives detonated when two of the bombs hit the ground, and plutonium dispersed across the Spanish village of Palomares.[48] Three bombs were immediately recovered, but one was lost at sea for 81 days, causing considerable embarrassment to the US.[49]

This was followed two years later by solar flare interference with NORAD sensors on 23 May 1967 – which was interpreted as a potential attack. On 9 November 1979, a computer error at NORAD resulted in alarm and preparations were made for a full-scale nuclear attack. On 15 March 1980, American warning sensors falsely determined that a Soviet missile was on its way to the US.[50] On 26 September 1983, Soviet early warning systems reported the launch of multiple ICBMs from bases in the US. Standing orders were to respond with an immediate and compulsory full-scale counter-attack. Only quick thinking by a single officer, Stanislav Petrov, prevented outright war. Even after the Cold War had ended, we wobbled close to the nuclear brink. On 25 January 1995, Russian radar systems misinterpreted a small rocket launched by a Norwegian research vessel as a ballistic missile, and Boris Yeltsin activated his nuclear briefcase.[51,52] As recently as 23 October 2010, US military officials at an Air Force base in Wyoming lost most forms of command, control, and security monitoring over 50 nuclear ICBMs for 45 minutes due to a hardware problem.[53] These are just a few of the most notable scenarios but there is a substantive body of peer-reviewed literature, and popular history books such as Eric Schlosser's 'Command and Control', that explore the history of nuclear near-misses in more detail.[54]

The Hawaiian False Missile Alert

The Hawaiian false missile alert has recently restored a feeling of nuclear dread among residents of the US and beyond.[55,56] In the summer of 2017, for the first time since the Cold War, Hawaii resumed monthly tests of its nuclear attack warning siren. This was followed by the announcement of a new public education campaign for nuclear attack preparedness in

November. Leaflets and multimedia announcements had already been distributed to the community, to provide guidance in the event of a nuclear attack from North Korea. Perception of the risk of nuclear war was already heightened among the community by these uncivil defence measures. On 13 January 2018, at 8:07 a.m., people in Hawaii received a message that read 'Ballistic missile threat inbound to Hawaii. Seek immediate shelter. This is not a drill.' This state of emergency lasted for 38 minutes, before the message was confirmed to have been sent in error, and was, in fact, a drill.[57] The Hawaiian false missile alert left the people of Hawaii fearing for their lives and anticipating an impending nuclear attack.

Hawaii has a 125-year long legacy of American occupation. Their monarchy was 'peacefully' overthrown in 1893 by American businessmen, and the US has proceeded to militarise this small cluster of islands, from the Second World War to the present day.[58] It has been difficult for indigenous communities to gain their rights to their authentic language and culture, and what remains has been turned into a pastiche for the fleeting pleasure of tourists. The people of Hawaii have survived this occupation though – and have developed a deeper insight into the meaning of living with risk. Hawaii is susceptible to tsunamis, flooding and volcanic eruption, and many residents already organise emergency kits for the hurricane season. Nobody is truly psychologically prepared for the advent of a nuclear attack. When we consider ourselves to be at risk, we mentally tally our limited choices and consider their outcomes. Most of us can only imagine how frightening it must have been to receive that short message on a Saturday morning. It was strengthened by radio and news broadcasts, which provided the impression that the cogs of a trustworthy and cohesive three-part warning system had been activated.

Hawaii was forced to contemplate the unthinkable for 38 minutes. Many of the thousands of people who received this accidental warning would have envisaged the irrevocable steps towards nuclear attack. They would have felt powerless, clutching their families and trying to decide how and where to shelter. They would have sent final messages to the people that they love. The trauma was real, even if the attack was not. It will take time for this community to recover and for the impacts to vulnerable people, families and children to fade. Nuclear anxiety is a rational response to a perceived threat, amplified by civil defence mistakes.

False alarms are dangerous, as they can cause the affected community to doubt the veracity of future warnings.[59] It was a wake-up call of the worst variety, forcing a community to prepare for death and destruction without warning. It demonstrated the need for the careful control of reliable emergency management systems, and for the provision of highly qualified emergency planners and managers in a state of constant and total vigilance. Or, alternatively, the need for improved international diplomacy and slow disarmament, to further diminish the likelihood of existential threats like nuclear war. That choice is in the hands of government.

INTENTIONAL THREATS

There are many ways that a cyber-attack could complicate the management of nuclear weapons, from attacks to C4 systems to false warnings, fake data and information blackout, with the potential for covert humanitarian aid data-frame hacking prior to nuclear attack. Without geographical technologies and systems, nuclear warfare is feeble. Geographers have a role to play in understanding this intersection – and the meaning of a hidden threat to a kinetic system.

Nuclear warfare is reliant on geotechnologies to function and has become increasingly vulnerable to cyber-attack as technologies have advanced. There is intersectionality between nuclear weapons, cyber-attack and geography. Cyber warfare has the capacity to affect the GIS and GPS systems used for nuclear defence.[60,61,62] The nuclear weapons system itself is part of a larger system for command, control, communications, computers, intelligence, surveillance and reconnaissance (C4ISR) that is heavily reliant upon cybersecurity and needs continuous protection from these digital threats. While lacking the immediately material consequences of traditional kinetic warfare, cyber-warfare creates an 'everywhere war' when it is used as a blunt tool to attack military systems. It has been described as cutting across the geostrategic domains for nuclear warfare, crossing 'land, sea, air and space'.[63]

The consequences of nuclear warfare are more profound for a digital society. It is not always possible to make a 'hard' back-up of geospatial data, stored in electronic databases on hard drives and other magnetic storage systems. The consequences of an electromagnetic pulse (EMP) produced by a high-altitude thermonuclear explosion are currently unexplored by geographers, although resilience against EMP has been considered in the context of storage and maintenance of 'hard' maps. Should electronic cartography be destroyed by nuclear warfare, society as we know it would be lost. Ironically, EMP is a specific risk for nuclear installations, which often don't use cloud-based storage for data, just in case of cyber-attack.

The possibility of nuclear terrorism has been of interest to politicians, scientists and academics since the 1980s. Terrorism does not comply with the laws of diplomacy, and defies the anticipated spatial, cultural and political geographies of nuclear warfare. While less catastrophic than the threat of nuclear warfare by state-sanctioned nuclear attack, concerns about nuclear terrorism are increasing, and its likelihood is greater.

There are four distinct nuclear terrorism scenarios. Nuclear terrorism could occur when a group or individual undertakes the theft of a nuclear weapon, or when a group or individual is given a weapon by a nuclear weapon possessor state. Alternately, in-situ nuclear installations, such as defence or energy facilities, could be attacked by a terrorist group. Or, in the most unlikely of these scenarios, terrorists could manufacture a nuclear device or 'dirty bomb' of their own.

The most effective way to prevent nuclear terrorism is by denying would-be terrorists access to nuclear materials. Despite attempts to secure post-Soviet nuclear sites, there are more than 1,800 metric tonnes of nuclear materials, including highly enriched uranium and plutonium, stored in hundreds of sites across 25 countries.[64] Some are poorly secured. There are global initiatives to counter this threat, including the 2016 Nuclear Security Summit.[65] This summit encouraged international and collective progress in improving nuclear security culture and best practice. The International Atomic Energy Association (IAEA) Convention on the Physical Protection of Nuclear Material (CPPNM), which first came into force on 8 February 1987, was renewed and revived at this event.[66] The CPPNM provides a legally binding international instrument for the control and management of nuclear material, by establishing measures to prevent, detect and punish radiological and nuclear crimes. There is also the Global Initiative to Combat Nuclear Terrorism, a global partnership of 86 nations and 4 official observers, who provide support against nuclear terrorism on a national and international level.[67]

Nuclear terrorism preparedness requires the identification of places that could be vulnerable to attack and understanding of the potential human responses and outcomes of such a scenario. The work of Parikh et al. has explored the geography of human behaviours after an improvised nuclear detonation or dirty bomb.[68] Their studies are geospatial, considering the movement and reactions of the public in the aftermath of a radiological or nuclear attack. Work has been done by the Department of the Environment in the US and Public Health England, the Defence Science and Technology Laboratory and Porton Down in the UK to understand nuclear terrorism risks. Further geographical research is needed to explore motivations, radicalisation and the locations at risk internationally, in the context of geopolitical and social risks.

Should the United Nations (UN) Ban Treaty be implemented, the risk of nuclear terrorism may increase if disarmament is not constantly, carefully and heavily regulated and monitored by external invigilators. Changes in the existing global pathways of nuclear products may create risks, and some states may illicitly seek to maintain the nuclear status quo. There is already a legacy of 'lost' nuclear materials since the post-Soviet denuclearisation of Ukraine, Belarus and Kazakhstan,[69] but perhaps we can learn from these mistakes. All fissile materials internationally could be geospatially tagged to create spatial databases that include isotope types and quantities, origin and provenance, and expected processing pathway. These databases could be shared with and verified by independent organisations, such as the IAEA, to create a complete forensic archive. There could be external spot-checks to ensure that materials are stored responsibly, but also to monitor the wellbeing, mental health and lifestyle choices of people with access to the material. Currently, these hypothetical safeguards would be impossible to implement due to issues of 'nuclear security'. However, if the UN Ban Treaty is successful, then it might make monitoring and surveillance – and accountability – more likely.

An under-examined and geographically interesting aspect of nuclear terrorism is insider threat risks. Materials go missing from nuclear installations on a surprisingly regular basis. There were 683 incidents of theft of nuclear materials internationally between 2013 and 2016, and these involved materials that would be suitable for radiological terrorism.[70] Theft of nuclear materials is the first stage of potential nuclear terrorism, and the crime is often undertaken by those inside the nuclear community, or with the support of insiders. Security clearance or defence-level vetting is needed for most roles inside these places, but this leads to a culture of complacency inside the nuclear military industrial establishment – suspects are seen as above reproach.

Evidence of crimes has emerged in the nuclear military industrial establishments. In the UK, nine Trident crew members were expelled after compulsory drugs tests revealed that they had taken cocaine in 2017.[71] A US Navy submarine commander faked his death to end an affair with his mistress in 2012.[72] A crewman of the UK's HMS *Vigilant* stole money from a sex worker in 2017.[73] On 8 April 2011, Able Seaman and submarine guard Ryan Samuel Donovan fired six shots from an SA80 rifle in the control room of the nuclear submarine HMS *Astute* in Southampton, UK. He shot dead Lt Cdr Ian Molyneux and injured a second crewman. This was at the same moment as local dignitaries, including Southampton City Council's mayor, chief executive, and leader, were being given a tour of HMS *Astute*.[74] This scenario could have been much worse. The military attracts those who are more likely to want to fight and take risks, those who need more excitement than civilians, and the process of becoming militarised requires that the value of life is demeaned to encourage men and women to fight. This means that the end-product of this process may be a human who is unfit for their role, but cleared to work on high-security projects. It does not take much imagination to anticipate that insiders could also steal nuclear materials, in addition to more banal crimes. Alarmingly, unlike other military miscreants and murderers, they do not always get caught. These are not the sort of people that we want guarding and maintaining our most lethal and dirty weapons.

BANNING THE BOMB

On 7 July 2017, the world changed when the UN Treaty on the Prohibition of Nuclear Weapons (Ban Treaty) was passed with a majority of 122 states.[75] This signalled to the world that nuclear weapons were unwelcome internationally, and that arms control was no longer adequate to prevent nuclear harm. If it is ratified by 50 governments, then the Ban Treaty could create a legally binding instrument to outlaw nuclear weapons. While nuclear weapon possessor states have declared that they intend to retain their arsenals, this is a big step towards a much larger nuclear-free zone.[76] This would at least restrict the parts of the world where nuclear weapons could be moved and deployed.

While the treaty was being crafted, the International Campaign to Abolish Nuclear Weapons (ICAN) declared that:

> We are on the cusp of a truly historic moment – when the international community declares, unambiguously, for the first time, that

nuclear weapons are not only immoral, but also illegal. There should be no doubt that the draft before us establishes a clear, categorical ban on the worst weapons of mass destruction.[77]

In 2017, ICAN were awarded a Nobel Peace Prize for their efforts. The Nobel Peace Prize has provided recognition and global attention for nuclear disarmament previously – honouring the IAEA in 2005, and Barack Obama in 2009. The Nobel Peace Prize Committee's choices have been noble – but sometimes naïve.[78]

There is already concern that the Ban Treaty is distracting from existing legislation, such as the NPT, which already encourages all signatories to pursue complete disarmament. The Ban Treaty, if successful, could create a deeper schism between non-nuclear and nuclear weapon possessor states as they choose to adhere to different treaties. If the Ban Treaty is successful, it is likely to resemble a large geographic nuclear-weapon-free zone (NWFZ), like those that already exist across the Global South. The original zones arose through local versions of these treaties, covering South America, the South Pacific, Southeast Asia, Africa and Central Asia. It is anticipated that the ban will hopefully unify existing NWFZs.

Nuclear weapon possessor states cannot be made to comply with any outcomes of the Ban Treaty, and the world is reliant upon their goodwill until the prohibition potentially becomes cemented in customary international law. It cannot create any legal necessity that could force non-signatories to disarm, as it only binds the signatory states under treaty law. The Ban Treaty could also have implications for NATO if members were to sign, leaving the alliance and making it harder to achieve constructive negotiation between Russia and the USA.[79] American experts fear that the Ban Treaty will create challenges for stationing and testing, and that it could undermine progress on IAEA safeguards.

The Ban Treaty simply cannot create a nuclear weapon-free world any time soon. However, it has the capacity to shake up existing approaches to arms control and may push forward existing non-proliferation and disarmament agendas. Over time, as more member states become party to the Ban Treaty, pressure for nuclear weapons possessor states to conform to the new norm will intensify. Nuclear weapons will become increasingly untenable, and will hopefully be consigned to history.[80]

8
Spaces of Peace

> Non-violence ... is the only thing that the atom bomb cannot destroy.
> – Mahatma Gandhi

Pacifism was described by sociologist Johan Galtung as any behaviours and actions that reduce the likelihood of short- and long-term violence, and that do not make a case for direct or indirect violence.[1] These actions are much-needed to prevent harm from nuclear weapons. The anti-nuclear pacifist movement arose at the same time as nuclear weapons, as scientists, academics, politicians and the public began to recognise the harm that was and could be caused by nuclear warfare. Anti-nuclear pacifism grew into an international movement, proliferated during the arms race, manifesting in networks, camps and nuclear-weapon-free zones (NWFZs). It has gradually changed the world's geopolitical dynamic by applying pressure for greater arms control and eventual disarmament.

Pacifist action can be found everywhere. From mass protests in New York, to non-violent action in Aldermaston, UK; from the European Union (EU) to the United Nations (UN), there are multitudes of places and spaces where people have felt empowered to express their dissatisfaction at the nuclear status quo. These activities serve as a counterbalance to the injustices and inequalities that have previously been described in this book, and as a conscience for the nuclear military industrial complex. Some activism is gentle and recurrent, such as the activities at Hiroshima Peace Day on 6 August every year. Other forms create semi-permanent safe spaces for activists. For instance, Greenham Common women's camp was in place at RAF Greenham Common in the UK for over 19 years from 1981 to 2000. From policy reforms to artistic expression, anti-nuclear pacifists have made the most of all resources available to them. However, the legacy of radical and game-changing activism by indigenous communities; Black, African and minority ethnic communities, and working-class women has been obscured, their places and spaces erased from history through neglect or misappropriation.

Activism against nuclear weapons does work, as evidenced by the implementation of nuclear-free zone international policy in the Global South. Geopolitical pushes are more likely to force nuclear weapon possessor states to adopt new paradigms, and these have often been instigated by shock. For instance, the 1963 Partial Test Ban Treaty, a global ban on atmospheric, extra-terrestrial and marine nuclear weapons tests was only ratified by the US, UK and USSR, following the Cuban missile crisis. It was more than thirty years later that all types of nuclear weapon testing were outlawed by the Comprehensive Test Ban Treaty (CTBT), after the Cold War ended.

Anti-nuclear pacifism defies the heteronormative and macho nature of nuclear weapons; it renounces the idea of deterrence through strength.[2] It reveals the devastating purpose of the bomb and challenges patriarchal discourse. It reveals the inherent misogyny that surrounds the nuclear military industrial establishment when the same terminology that is often used to trivialise and disempower women is deployed on anti-nuclear pacifists.[3] However, anti-nuclear pacifism is not ignorant and emotional. It is not irrational or irresponsible. It is not attention-seeking or divisive. When faced with the sinister and murderous truth of nuclear warfare, arguments of pragmatism or necessity have little rational countenance.

Anti-nuclear activism holds the nuclear military industrial complex to account for obscuring the links between nuclear energy and nuclear defence. Anti-nuclear activism reveals the networks of injustice created by the nuclear sector in the name of globalisation. It publicly exposes the tragedies of historically militarised atomic atolls, as a concentrated synecdoche for the international consequences of nuclear deterrence. Anti-nuclear activism is not irrational. Instead, it provides a sensible discourse towards pacifism, and a counterpart to the consequences of a heedlessly militarised planet.[4]

ANTI-NUCLEAR PACIFISM

Anti-nuclear pacifist action occurred before the atomic bomb was dropped on Japan, and the first people to become aware of the dangers of nuclear warfare were the atomic scientists themselves. The Szilárd Petition was signed by a committee of 68 Manhattan Project scientists who were in moral opposition to the atomic bomb, and was delivered to President Truman in July 1945.[5] He ignored their pleas and proceeded to bomb Hiroshima and Nagasaki. Many more scientists, including Oppenheimer himself, became staunch opponents of nuclear warfare after the devastation of the bombings was revealed.[6]

In 1946, eight of the Szilárd Petition signatories coalesced to become the Emergency Committee of Atomic Scientists (ECAS).[7] This group of anti-nuclear activists was headed by Albert Einstein and Leó Szilárd, who were determined to warn and inform the public of the dangers of the atomic bomb.[8] Half of the members of ECAS had worked directly on the Manhattan Project, and all of them had been indirectly involved in the production of the first atomic bomb. The critical theorist Lewis Mumford was among those who became pacifists after the bombs fell, noting that: 'Not the least extraordinary fact about the post-war period is that mass extermination has awakened so little moral protest.'[9] As often happens in these situations, the experts were ignored, so that a geopolitical agenda of dominance could be pursued instead. However, by 1962, Linus Pauling won the Nobel Peace Prize for his work to stop the atmospheric testing of nuclear weapons, and the 'Ban the Bomb' movement had spread throughout the United States.

Across the Atlantic, Polish physicist Sir Joseph Rotblat formed the Atomic Scientist's Association (ASA) in the same year.[10] The ASA was a politically neutral group with similar aims to ECAS, including the promotion of international control of nuclear issues. Over the next decade, many intellectuals refashioned their professional identities, calibrated their moral compasses and formed non-governmental organisations in science and medicine to prevent nuclear war and weapons testing.[11] These organisations heralded a new move towards non-violent internationalism.

The ASA gained prominence when the Russell–Einstein pacifist manifesto was issued on 9 July 1955.[12] This high-profile manifesto called for world leaders to find peaceful resolutions to conflict instead of pursuing nuclear proliferation and was signed by eleven eminent experts including Joseph Rotblat, Max Born and Albert Einstein. The intellectual anti-nuclear activism movement grew when Rotblat collaborated with British mathematician Bertrand Russell to form the Pugwash Conferences on Science and World Affairs in Pugwash, Canada in 1957.[13] Twenty-two scientists attended the first Pugwash event, with many countries, from the USA to Japan and the Soviet Union, represented. Pugwash showed that intellectual anti-nuclear activism was an international movement, transcending the colonialist and anti-communist sentiments that were sometimes espoused by American and British anti-nuclear activists.[14] New expert groups emerged, with the most significant being the Nobel Prize-winning International Physicians for the Prevention of Nuclear War (IPPNW) network.[15] This group called for multilateral disarmament and criticised British civil defence, causing consternation for Prime Minister Margaret Thatcher. It succeeded in opening channels of transnational communication during the late stages of the Cold War. The success of the IPPNW in the 1980s helped to precipitate the creation of the International Campaign to Abolish Nuclear Weapons (ICAN) in the 2010s.

A professional transnational anti-nuclear pacifism network that has survived to this day is the Bulletin of the Atomic Scientists (BAS), which formed due to recognition of a need for greater availability of public information in 1945.[16] BAS conveys information about nuclear weapons in a non-technical way, to try to ensure accessibility to policy makers, journalists and the public. One of the ways that they communicate risk is through their Doomsday Clock, which was set at seven minutes to midnight when it was devised in 1947. There is some controversy over the Doomsday Clock, as its original setting was designed for aesthetic interest rather than being set by any specific scale or deadline. It is reset annually, and has become a useful indicator of changes in nuclear risk. In 2018, it was reset at two minutes to midnight, the closest it has ever been set to annihilation. This is because the existential risk of climate change has been included in its battalion of concerns.

WOMEN AND PEACE

While women were only given back-seat roles in the Manhattan Project, they came to the forefront of early anti-nuclear activism. Women Strike for Peace grew from a protest held against atmospheric nuclear testing on 1 November 1961. This protest brought together approximately 50,000 women, who marched across 60 cities in the USA to demonstrate against nuclear weapons. It was the largest women's national peace protest of the twentieth century. This group voiced concerns about harm to women's and children's health, and the future of the planet. They picketed the White House and the Pentagon and condemned the Japanese bombing in conjunction with the Women's International League for Peace and Freedom. Peace camps sprang up, including, in the 1980s, Greenham Common in the UK.

The space of the peace camp is interesting from a geographical perspective, as it offers a specific and exclusive space where the status quo of nuclear deterrence is contested, a place where societal norms can be subverted and an immersive counter-culture can be established. A notable feature of peace camps in the UK is their gendered nature, which demonstrates the power of women's activism and the creation of safe spaces for peace by and for women. One of the most significant of these camps is the Greenham Common Women's Peace Camp, which was established at RAF Greenham Common in Berkshire, UK. The camp originated in September 1981 when a group called Women for Life arrived at Greenham to protest cruise missiles. The first camp blockade occurred in May 1982, with 250 women protesting, during which 34 arrests were made. The camp evolved from a temporary fixture into a permanent location and was active as a women-only camp for 19 years, before finally disbanding in 2000. It became a symbol of women's resistance against the male-dominated world of nuclear weapons. Greenham Woman Sarah Hopkins recalled in her memoir that the RAF base had become 'a symbol for nuclear error, male domination and imperialist exploitation'.[17]

Greenham Common gave women their own space to peacefully protest the bomb, where the few men present were assigned traditionally female jobs of childcare and domestic labour. The camp was closely linked to both the CND (Campaign for Nuclear Disarmament) and the wider women's movement, but independent of both. It became known not as a singular protest, but rather a point of interaction between various protest movements. Initially, there was opposition by the CND and wider peace movement that men were barred from the camp, while the women viewed the CND as condescending and patriarchal, respectable and establishment. This distrust gradually dissipated as the determination, energy and creativity of the Greenham Women became clear. They never gave up. During the time that this camp was active, the women camped outside, breached the perimeter fence, faced criminal charges and disseminated a transnational message of nuclear disarmament. On 12 December 1982, 30,000 women encircled the RAF base, representing the most effective single action undertaken by the peace movement in the 1980s. Geographer Tim Cresswell has explored their subversion and transgression of geographical gender norms, describing their work as a 'carnivalesque protest'.[18]

Other notable camps are the Molesworth People's Peace Camp, which was established in December 1981 and included men. This camp exhibited different values and strategies to Greenham Common, with a Quaker presence at the camp, supplemented by people of other faiths and various leftist and counter-cultural persuasions. The camp became a link in a Europe-wide network of centres for non-violent direct action in opposition to NATO plans to deploy Pershing II and Ground Launched Cruise Missiles. One aim of occupation of the site was to goad the Conservative government. They claimed that they could dismantle the fence by night faster than it could be erected by day, gaining publicity for the anti-nuclear cause. Despite deliberately provocative activity by camp residents there was little in the way of direct confrontation between protesters and military personnel. Other UK camps include the Faslane Peace Camp in Scotland, the longest running peace camp in the world. Dr Catherine Eschle has explored how the 'everyday' is reproduced there, as the camp community lives non-violently and challenges economic norms.[19]

AFRICAN-AMERICAN PEACE

In the story of every incredible achievement, women and minority communities are pushed aside to tell only the story of white men. This is true of the historic contributions made by African-American people to the American anti-nuclear pacifist movement.

The first public demonstrations against nuclear weapons testing were undertaken in Times Square, New York, in July 1946. The movement was formed following the South Pacific nuclear weapons tests. According to Paul Boyer, 'it was Bikini, rather than Hiroshima and Nagasaki, which first brought the issue of radioactivity compellingly to the nation's consciousness'.[20] American peace activism since the late 1940s has focused on opposition to American foreign policy, and support for nuclear non-proliferation and arms control. Protesters in the US had to face the American war machine and the draft during the Vietnam years, but peace activism still flourished, influencing both domestic politics and international relations, and helping to shape a peculiarly American culture of radicalism. Some of the more significant outcomes of the American anti-nuclear movement included the 1 million people who attended the June 1982 New York City Rally, and the equally prominent Great Peace March for Global Nuclear Disarmament in 1986.

Campaigns by international pacifist organisations including Ground Zero and Pacific Life instigated some unusual tactics, committed to non-violence in the tradition of Gandhi and Martin Luther King, Jr. For instance, on 6 July 1975, when 30 people crossed the fence of Puget Sound Naval Base with gardening supplies and proceeded to plant a vegetable garden before being evicted.[21] Blockades were also organised to prevent new nuclear submarines from docking in 1982, by members of Ground Zero. Much anti-nuclear activism of this era was overshadowed by the violent events of the Vietnam War, so high-profile non-violence became the dominant tactic and trademark of American anti-nuclear activism.

American peace activism and the African-American civil rights movement are uniquely and deeply connected. It was quickly realised that the rise of nuclear weapons was a threat to the civil rights of everyone. Nuclear weapons are not only an inherently racial and colonial issue, but they are also expensive. Every dollar spent building more and more powerful bombs could have been used to support welfare and equality in society.

Members of the African-American community recognised these issues and played a significant role in anti-nuclear activism, being among some of the first citizens to protest Truman's decision to drop atomic bombs on Hiroshima and Nagasaki in 1945.[22] Black liberals, artists, clergy, musicians and civil rights leaders were involved in activism, including Dr Martin Luther King, who often spoke directly of the dangers and challenges of the bomb. King is quoted as saying:

> Somehow, we must transform the dynamics of the world power struggle from the negative nuclear arms race, which no one can win, to a positive contest to harness man's creative genius for the purpose of making peace and prosperity a reality for all of the nations of the world. In short, we must shift the arms race into a 'peace race.' If we have the will and determination to mount such a peace offensive, we will unlock hitherto tightly sealed doors and transform our imminent cosmic elegy into a psalm of creative fulfilment.[23]

The colonial consequences were evident to these communities. Vincent Intondi describes this in his excellent exploration of Black anti-nuclear activism: 'For black leftists in Popular Front groups, the nuclear issue was connected to colonialism: the U.S. obtained uranium from the Belgian controlled Congo and the French tested their nuclear weapons in the Sahara.' This Black activism was transnational and global in reach. Beyond the USA, Black peace activists travelled from newly independent Ghana to the French Saharan nuclear weapons test sites. This was a significant movement against nuclear colonialism, and a counter-narrative to the disempowerment of scholarship on Africa since the 1980s.

JAPAN AND PEACE

As the only place where nuclear warfare has been intentionally used, Japan had an inherently pacifist approach to the Cold War arms race. The bombings of Hiroshima and Nagasaki were seared into public memory. The fallout from these attacks meant that the Japanese public were considerably more aware of the issues arising from nuclear weapons. However, there is a further incident of consequence to Japan. The *Lucky Dragon no. 5* was a fishing ship with 23 crew, fishing in the Midway Sea for tuna. On 1 March 1954, their course took them close to Bikini Atoll, through the fallout from Castle Bravo, the largest bomb ever tested by the US.[24] The crew had to be treated for acute radiation syndrome (ARS), and many were then inadvertently infected with hepatitis C through blood transfusions.

Hibakusha anti-nuclear activist Muto Ichiyo argued that the rhetoric of peaceful use of nuclear energy sought to turn 'the personal experiences of individual [atomic bomb victims]...into an abstraction with no individual faces or bodies'.[25] In the midst of anti-nuclear sentiments spurred on by the *Lucky Dragon* incident, the US and Japanese governments launched a 'peaceful' nuclear energy program, with an aim to minimise anti-nuclear activism by remaking the image of deadly weapons. Its aim was to recast nuclear technology as prosperous and modern, instead of lethal. This led to a delay in the publication of *Hibakusha* testimonies, in favour of pro-nuclear energy propaganda. The *Hibakusha* resisted nuclear energy in the 1970s, but a pattern of denial and cover-up by industry and government helped create a dichotomy between nuclear energy and nuclear power, an illusion of a distinction between them that occurred not just in Japan but internationally. This discourse was temporarily disrupted by the Fukushima nuclear power plant disaster of 2011, with low-level and high-level anti-nuclear energy activism arising. Japan is a relatively pacifist non-nuclear weapon possessor state, and the harmful effects of the atomic bomb are memorialised in beautiful peace memorial parks, and shared globally though the stories of the *Hibakusha*.[26]

RUSSIA AND PEACE

Despite the Soviet Union being a nuclear weapon possessor state, the historic culture of Russian anti-nuclear activism has been heavily influenced by official communist pacifist organisations like the World Peace Council (WPC).

The American government was deeply concerned by Soviet pacifist activity during the Cold War. In 1951, the House Committee on Un-American Activities published *The Communist 'peace' offensive*, which detailed the activities of the WPC and of numerous affiliated organisations and people, who were suspected of being 'communists', due to their involvement in peace movements, treaties and events. In 1982, the Heritage Foundation published *Moscow and the peace offensive*, which claimed that non-aligned peace organisations advocated similar policies on defence and disarmament to the Soviet Union. It argued that 'pacifists and concerned Christians had been drawn into the Communist campaign largely unaware of its real sponsorship'. *Time* magazine even suggested that the scientific nuclear winter hypothesis was propagated by Moscow to give anti-nuclear groups in the USA and Europe fresh ammunition against American nuclear arms proliferation.[27] Astonishingly, the Soviet Union's Main Intelligence Directorate (GRU) really did provide funding to much of the American pacifist movement, and supported their demonstrations in Europe against American nuclear bases.

It is notable that rather than representing anti-nuclear convictions, the anti-nuclear protests in the Soviet Union during the late Perestroika period (1985–87) reflected an explosion of anger and resentment against Moscow's domination. For some, anti-nuclear activism became a surrogate for nationalism, and a way to oppose Moscow's imperial treatment of the republics. For others, it was a front for anti-communist protest and a way to demand greater local self-determination.[28]

However, there was less-politicised academic recognition in the Soviet Union of the dangers of nuclear warfare. In 'The geography of peace and war: A Soviet view', published in 1985, Gerasimov[29] noted the major task for political geography in Soviet pacifism. He issued his own 'Geographers for peace, against the arms race and the nuclear war threat' call to arms, specifically aimed at geographers of Soviet nations. Within this call, he specified that:

> We have to disclose the truly global after-effects of a nuclear catastrophe, including those 'limited' nuclear conflicts, to show that in modern war there will be no victors and no vanquished, that the population of neutral countries will also be doomed to perish – even those situated many thousands of kilometres from battlefields.

He notes that 'Overall disarmament would improve the international climate, would contribute to scientific and economic collaboration in solving problems which concern all people: those of fighting hunger and maladies ... of using marine resources and space research for peaceful purposes.' He called for all geographers to use their 'professional knowledge and potential for strengthening peace among the nations'. This message echoes that of radical Western geographers of the era, although it is debatable how much of it is state-constructed anti-nuclear propaganda, versus genuine anti-nuclear sentiment.

Since the end of the Cold War, the Russian cultural approach to nuclear weapons has become more divisive and more distinct from current Western practice. There is reportedly a more general acceptance of nuclear deterrence, and little evidence of anti-nuclear pacifist action being condoned by the state. Pacifism has not been a topic for schools in this militaristic state since the early 2000s. Anti-nuclear activists are labelled as ecoterrorist subversives. They face heavy penalties and are even imprisoned or exiled.

ATOMIC ED AND THE APOCALYPSE

Los Alamos has long been a place of anti-nuclear activism. From protests to hunger strikes, both local and international communities have gathered here in non-violent protest against nuclear weapons. Much activity has also been undertaken by a working group to address Los Alamos health concerns.

Ed Grothus created a permanent enclave for pacifist and anti-nuclear activity in Los Alamos, a hol(e)y space of activism, within his radically pacifist Church of High Technology. In the 1950s, Grothus was one of the machinists who helped to manufacture the bombs at Los Alamos Nuclear Laboratory (LANL). However, disillusioned with the US war in

Vietnam, Grothus became an anti-war activist, and decided to fund his new life of pacifism by selling cast-offs from the nearby nuclear defence lab in a repurposed mom-and-pop store, that he called the Black Hole of Los Alamos. Upon entry, visitors were greeted by a sign that said 'One bomb is too many. No one is secure unless everyone is secure. Don't throw anything away. Welcome to the Black Hole'.

The Black Hole was a world class atomic junkyard emporium, a perpetual nuclear yard sale with artefacts spanning the entire history of LANL. He arranged his treasures as a museum of sorts, a disorganised meta-inventory of discarded atomic objects. Grothus hoarded obsolete seismographs, oscillographs, spectrometers and detonation cables, alongside bomb casings and filing cabinets of scientific notes. A mishmash of atomic technologies and trinkets nestled incongruously beneath peace signs. His vast, mouldering collection and peacenik activities made Grothus something of an embarrassment to the local community. The Black Hole became of concern to the FBI after Grothus sent a can of sweetcorn that had been relabelled as 'Organic Plutonium' to the White House. He also had a habit of labelling any old hard drives that were laying around in his Black Hole with 'Secret' stickers, regardless of content, and these were subsequently confiscated by the FBI.

Grothus supported the work of surrounding pacifists and artists, among them Tony Price, who created atomic appliances and appendages, producing Strangelovian montages with an anti-nuclear ethos. He used the scrap from the Black Hole and LANL as components for monstrous metallic installations, with titles including 'Native Who Sold His Island for A Nuclear Test' and 'Post-Apocalyptic Conference of Metallic Diplomats'.

Grothus purchased an old Lutheran church next to the Black Hole in 1966, initially as a storage space for his nuclear knick-knacks. However, it was soon transformed into a something that was almost theological, a sheltered platform for the most subversive performance art of Los Alamos. It became the Church of High Technology, a place to go and 'unworship the bomb', in the heart of the atomic city. Grothus became a self-ordained cardinal, calling himself 'Don Eduardo de Los Alamos', and each Sunday, he would dress up in bishop's robes and preach the 'critical mass'. He gave sermons in the role of cardinal, using a mantle of moral authority to work towards peace and against nuclear weapons. Fire and brimstone had nothing on Grothus; he made doomsday predictions, warning of nuclear apocalypse, that became more extreme as his life progressed: 'I predict it will happen in 2013, when an American with an American weapon of mass destruction destroys Washington, DC, which starts a nuclear holocaust and everyone on Earth dies.'[30] He was the Los Alamos harbinger of nuclear dread. However, he needed something more permanent to remind everyone at LANL of the risks of nuclear defence.

Ed's final project was to order, from China, twin 40-ton white granite obelisks as monuments to the atomic age, his own 'Doomsday Stones': '[They] are not to celebrate the bomb but to make note of the most important man-caused event in the history of the world', he explained, describing them as Rosetta stones for a nuclear age.[31] They were shipped from China, inscribed in 15 languages with warnings of the consequences and risks of nuclear weapons. Two massive hunks of black granite, salvaged from LANL seismologists, were intended for use as plinths for these gigantic monuments. These hunks of rock lingered in the yard of the Black Hole for some time before Grothus came to the idea for the monument. His vision would be crowned by two black granite spheres, with a design that replicates the charges that surround the core of a nuclear weapon. Grothus was inspired by the ancient Egyptian obelisks, massive pillars inscribed with information for travellers that stood at the entrance of towns. He wanted to create the Los Alamos obelisks as a warning to all who entered the atomic city. The inscription on the pillars is:

> Welcome to Los Alamos, New Mexico, the United States of America, the city of fire. Our fires are brighter than a thousand suns. It was once believed that only god could destroy the world, but scientists working in Los Alamos first harnessed the power of the atom. The power released through fission and fusion gives many men the ability to commence the destruction of all life on Earth. Nuclear bombs cannot be used rationally and dreams for safe and useful nuclear power may never be realised. It is only in Los Alamos that the potentials for unimagined, fantastic good and demonstrated, horrendous evil are proximate.

While the Doomsday Stones were not erected within Grothus' lifetime, they did have an outing on 6 August 2005, the fiftieth anniversary of the Hiroshima bombing. On this occasion, he loaded a Doomsday Stone and several buckets of sunflowers onto a large trailer, latched to his red pick-up truck. After donning his cardinal's garb, he drove slowly across Los Alamos until he reached the park before unloading the trailer in front of a rapidly growing crowd. As the crowds thinned, he reloaded his truck, and climbed into its cabin before falling asleep. He hoped eventually to erect the obelisks in front of the Black Hole, if nowhere else. The Doomsday Stones currently remain in several large steel containers behind the Black Hole, rather than being publicly visible. His daughter, Barbara Grothus, said that there had been challenges in gaining permission to erect the Doomsday Stones in such proximity to Los Alamos. Grothus wanted to donate his monuments to Los Alamos County, but the local community voted unanimously to decline his offer. Ed died in 2009, and the Black Hole was eventually closed in 2014. It was the end of a monument to his country's history, and his unique way of keeping those memories alive.

The Church of High Technology is currently derelict, with smashed windows and graffiti smothering the exterior walls. The Black Hole has become a place for a different type of counter-culture, as a space of delinquency within an otherwise stringently regulated environment. The children of LANL workers creep down here, to graffiti, throw bricks at windows and smoke weed behind the church. Several signs are emblazoned across the back of the church, warning away trespassers and threatening to introduce hidden animal traps to the area. The inside of the church is dishevelled and run-down. The rafters are festooned with messy birds' nests, and mouse droppings litter the floors surrounding the pulpit. Where a Bible should be on the pulpit, is instead a tattered copy of *Energy Information Abstracts*. Despite Grothus' departure, the Church of High Technology remains a place for the disenfranchised of Los Alamos.

PEACEFUL POLITICS

Two of the most notable organisations that have promoted international nuclear disarmament are the WPC,[32] a communist pacifist organisation founded in 1950, emerging from the Soviet Union policy to promote international peace campaigns and oppose 'warmongering' from the USA, and the more westernised British CND.[33]

The WPC's first conference was undertaken in Sheffield and Warsaw. It was denounced by Prime Minister Clement Attlee as 'bogus forum of peace with the real aim of sabotaging national defence'. A limit was imposed on foreign delegates, reducing participation from 2,000 to 500 delegates. This led to the exclusion of WPC founder Frédéric Joliot-Curie and the Soviet pacifists Ilya Ehrenburg, Alexander Fadeyev and Dmitri Shostakovich. By March 1950, the WPC had released the Stockholm Appeal, calling for an absolute ban on nuclear weapons. The WPC had several notable supporters, many of whom were communists, among them Jean-Paul Sartre, Pablo Picasso, Paul Robeson and Diego Rivera.

In 1955, the same year as the Russell–Einstein manifesto, the WPC launched an 'Appeal against the Preparations for Nuclear War'. The WPC organised international congresses with pro-communist organisations and states until the late 1980s. Large-scale funding was withdrawn in 1991 following the collapse of the Soviet Union. It continues to run conferences and attract delegates from nuclear-free and former Eastern bloc countries, but it operates within a less high-profile domain now. It describes itself as an:

> anti-imperialist, democratic, independent and non-aligned international movement of mass action. It is an integral part of the world peace movement and acts in cooperation with other international and national movements. The WPC is the largest International Peace structure, based in more than 100 countries.

The CND was 60 years old in 2018. It has continued to gain prominence as a transnational anti-nuclear movement. The origins of the CND are multifarious,[34] but its formation is attributed to opposition to the nuclear weapons tests from Labour Party MPs in 1954, as well as from other quarters including National Christian Council of Japan, the Quakers and the British Council of Churches.

By the 1960s, newer non-politically aligned organisations were contributing to an East–West pacifist divide. Pacifist activists Bertrand Russell and Spike Milligan formed an Emergency Committee for Direct Action against Nuclear War, encouraging *direct action* and civil resistance against the tests. This meant that they encouraged sit-ins, strikes, workplace occupations, blockades, protests against the bomb, including the actions of the Quakers Sheila and Harold Steele, who decided to try to sail a boat into the test area to raise awareness and prevent the testing going ahead.

In April 1958, a march took place organised by the Direct Action Committee against Nuclear War (DAC) and the CND. Several thousand people marched from Trafalgar Square in London to the Atomic

Weapons Research Establishment in Aldermaston, to demonstrate their opposition to nuclear weapons. This was the first of many CND marches, camps, protests, human chains and other forms of direct action over the decades. The CND initially espoused unilateral nuclear disarmament for the UK, alongside global abolition. Its first wave of activity occurred from 1957 to 1963, followed by revivals in 1980 to 1990, and a more recent and unsuccessful flurry of activity surrounding the vote for potential abolition of the UK Trident missile system in 2016.

Given the singular nature of the movement, the campaign initially included many disparate characters. The metropolitan and middle-class Golders Green Guildswomen were instrumental in the formation of the CND, raising awareness of the issues surrounding nuclear weapons and aligning sympathetic political and faith groups. The early CND executive meetings were also notable for their inclusion of prominent women, including academics, authors and political activists. Although inclusion should not be conflated with influence, there was power in numbers. An early pamphlet for the CND posited an 'Appeal for Women', urging responsible women to work towards a better world for their children and arguing that the funding spent on nuclear defence could instead be channelled into humanitarian issues. Hence, the CND used women's symbolic capital to encourage political sympathisers such as the Labour Party of the era, to contemplate unilateral disarmament of the UK. Comically, CND campaigner Freda Ehlers is quoted as saying: 'In CND I have to mix with so many odd people, the sooner we ban the bomb the better.'[35]

The Committee of 100 grew from the CND in 1960, starting with 100 public signatories supporting mass non-violent resistance and civil disobedience to campaign against nuclear weapons. Among those signatories was Bertrand Russell, who was convicted of inciting civil disobedience in September 1961 and imprisoned for a week, at the age of 90. The Committee of 100 began to organise mass sit-down demonstrations of more than 2,000 people. Their first act of civil disobedience occurred on 18 February 1961 outside the Ministry of Defence, to coincide with the arrival of USS *Proteus* on the River Clyde. Further similar demonstrations were undertaken in Parliament Square, the US and Soviet embassies in London, and at the Holy Loch Polaris submarine base. Things escalated from there, and protests began to number in the tens of thousands and result in arrests. Members of the Committee of 100 were also responsible for the Spies for Peace activities in 1963, whereby a small group of peace activists broke into a secret government nuclear bunker and stole secrets about Regional Seats of Government activities in the event of a nuclear war. This course of action would be impossible in today's high-surveillance society.

Nowadays, the CND take a gentler approach to their activism, from organising sit-ins outside nuclear installations to celebrating Hiroshima Peace Day in Southampton, UK. The CND collaborates with organisations like ICAN to raise awareness of the need for disarmament.

PACIFISM NOW

While the Global North was stockpiling nuclear weapons, the Global South had other ideas. The people of the South Pacific, South America and Africa recognised the dangers of nuclear proliferation and wanted a way to declare themselves out of the game. They collaborated to create the multilateral Treaty of Rarotonga, which declared the South Pacific as a new nuclear-free zone. This treaty recognised the harm that had been done to communities in the South Pacific due to nuclear weapons testing.

The treaty was a long time coming, after considerable discontent from many nations, notably those in Africa and the South Pacific, where nuclear weapons tests were carried out in a surge of nuclear imperialism by the US, UK and France. Ghana's first prime minister and president, Kwame Nkrumah, remarked that 'the poisonous fall-out did not, and never will, respect the arbitrary and artificial divisions forged by colonialism across our beloved continent'. He declared that 'Africa is not interested in such "defence" which means no more than the ability to share in the honour of destroying mankind.'[36]

This transnational pacifist movement was supported by international treaties and strong pacifist governments, including that of New Zealand. In 1984, the Prime Minister of New Zealand, David Lange, created a large nuclear-free zone, before the treaty was implemented, by banning nuclear-powered or armed ships from entering New Zealand's waters. This was the first stage of a New Zealand Nuclear Free Zone, Disarmament and Arms Control Act of 1987. This act also prohibits New Zealand residents and citizens from having the capacity to 'manufacture, acquire, possess, or have any control over any nuclear explosive device'. This demonstrated the strength of national-level policy for creating positive peace and restricting the global networks of nuclear deterrence. The act remains a component of New Zealand's international foreign policy, and the only permitted uses of radioactive materials are for peaceful medicine, research and energy purposes.

This has been taken one step further by the Nuclear Ban treaty, where 122 nations voted for prohibition of the development, production, possession, testing, use and threat of use of nuclear weapons. The treaty remains open for further ratification signatories at the UN. The treaty is not yet supported by any nuclear weapon possessor state, but it is understood that this decisive transnational action has taken steps to re-stigmatise nuclear weapons, re-establish the nuclear taboo and reinvigorate public debate and activist action for the complete abolition of nuclear weapons.

Almost fifty years since the implementation of the first Non-Proliferation Treaty (NPT), anti-nuclear activism and the concept of nuclear weapons elimination has evolved from a dangerous, utopian idea to a goal that is regarded as desirable and feasible by most of the world. The International Court of Justice declared in 1996 that 'in view of the unique characteristics of nuclear weapons', their use 'seems scarcely reconcilable with respect' to the prohibitions of methods and means of warfare 'which would preclude any distinction between civilian and military targets'.[37] The long-lived radiation released by nuclear weapons cannot be controlled in space and time and is therefore not limited to military practices and exposures. Protection against the effects of nuclear weapons is not feasible. The challenge now is to persuade nuclear weapon possessor states that disarmament is a justifiable and sensible pathway to take.

The ICAN organisation has played a significant role in pacifist progress in the 2010s. Malaysian obstetrician and former co-president of IPPNW Ron McCoy first proposed the idea of ICAN in 2005.[38] ICAN have adopted a discursive strategy, borrowing from critical and post-positivist international relations (IR) theories.[39] They have tried to contest the dominance of national security narratives and force nuclear weapon possessor states to address the long-term humanitarian consequences of nuclear weapons.[40]

The Asia-Pacific Director of ICAN and UN Ban Treaty coordinator Tim Wright is out there, eloquently sharing his organisation's understanding of nuclear harm. He returned to Japan in July 2018, to continue his rationalist argument, saying:

In time, nuclear-armed states, and the allies whom they 'protect', will come to realise the wrongness of their policies and actions. They will join the international mainstream in opposing nuclear weapons. Any leader who abhors the use of chemical weapons, who cannot tolerate the sight of injured and dying children in Syria or elsewhere, must recognize that nuclear weapons are an even greater menace requiring nothing less than our full-throated opposition.[41]

It is difficult for any humanitarian to disagree with this powerful and relational argument, yet it has still not been possible to persuade nuclear weapons possessor states to disarm.

ICAN originated in Australia, but the Australian government has turned its back on ICAN's policy successes and played no part in shaping the initial draft of the landmark treaty.[42] Despite being a signatory of the NPT and the South Pacific Nuclear Free Zone Treaty, Australia is reluctant to relinquish its perceived 'protection' of the US nuclear umbrella, tacitly endorsing extended nuclear deterrence. Put simply, Australia wants the benefits of protection from a nuclear weapon possessor state, but would rather nuclear warfare happened a little further away. This is understandable, considering the contamination and trauma, especially for Aboriginal people, after the UK's Maralinga and Montebello Island tests.

This legacy could be a motivation for Australia not to engage in the Ban Treaty, as it stipulates that 'state parties in a position to do so should provide assistance – including medical care, rehabilitation and psychological support – to individuals affected by the use and testing of nuclear weapons in areas under their jurisdiction or control'.[43] This could mean that genuine reparations would have to be provided to everyone who suffered, at the Australian's government's expense.

Anti-nuclear activism is ongoing, and it is not always peaceful and dignified. A recent example of more radical and criminal behaviour is the actions of an international group of five peace activists who invaded the German Büchel Airbase on 17 July 2017.[44] They cut two external fences and reached the bunker that stores American nuclear weapons, where they sat for an hour and scrawled 'disarm' on the bunker door. This activity triggered an alarm, and the group were arrested two hours after entering the base. The group said: 'we invaded the Büchel air base non-violently to capture the nuclear weapons stored here'. We ask Germany either to dispose of the weapons or to send them back to the United States to be disarmed there.'[45] After an hour in custody, where they were searched and photographed, the five were released through the main entrance to the military base.[46] This highlights the concerning security risks and breaches that surround nuclear weapon bases. Ironically, security on these sites often consider the behaviour of activists to be the equivalent of penetration testing – a harmless breach of security that enables them to identify and fix vulnerabilities.

Anti-nuclear activism can be considered a pathway towards the future abolition of nuclear weapons. It is not a static endpoint, and many steps towards nuclear pacifism have been taken during times of heightened risk of nuclear conflict, such as the Cold War. Creating a nuclear weapon-free world is a precarious socio-spatial process, shaped by the places where peace is made, such as the NMFZs worldwide. Positive nuclear peace is not simply an absence of nuclear warfare or military pacification. It is the complete abolition of nuclear warfare, and even its possibility in the future. Nuclear peace should be considered in the context of social justice, development and human rights, not just warfare. Geography poses a specific conceptual problem for anti-nuclear activism, as it must simultaneously operate beyond conventional forms of territorial politics, while remaining politically flexible for strength and relevance.[47]

Often activist action has been catalysed by geopolitical instability and characterised by the challenges surrounding the leadership of several nuclear weapon possessor states. Activists persist in raising awareness of the dangers of nuclear warfare, the wasteful expenditure of maintaining a deterrence, the human and environmental harm that originate from the nuclear military industrial complex. Nuclear weapon possessor states continue to depict their activists as irrational and uneducated, despite a legacy of activist networks that include experts, intellectuals and academics of all genders and ethnicities. However, major changes have arisen due to nuclear activism, precipitated by geopolitical shocks such as the Cuban missile crisis. The multilateral risks that have emerged since 2002 from North Korea, the Middle East and Pakistan are catalysing this fresh change.

9
Future War Zones

> We must abolish nuclear weapons, or they will abolish us.
> – John F. Kennedy

The world is heading towards a new era of nuclear risk, and perhaps a second Cold War. Unfortunately, existing international treaties seem powerless to prevent concerning changes in the current nuclear status quo. Unlike biological and chemical warfare, nuclear warfare is the only weapon of mass destruction (WMD) that has not yet been internationally banned. International arms control and non-proliferation treaties are the only significant limiting policies in place to prevent nuclear war from breaking out. The paradox, of course, is that the number of nuclear weapons has decreased –from a peak of 70,300 during the Cold War to 14,200 in 2018[1] – but the number of nations that possess nuclear weapons, and the overall threat of nuclear warfare and terrorism has increased. After years of relatively stable relations between nuclear-armed states, there is a heightened and increasing state-level threat.[2] However, our future risks and conflicts remain bounded and determined by geography and geopolitics.

The future of nuclear warfare is still unwritten, but the world seems stuck in an era of negative peace. Existing arms control measures are inadequate, and state-led ones have been quashed by short-sighted governments. For example, on 25 January 2017, two Democratic US Congressmen unsuccessfully attempted to introduce the Restricting First Use of Nuclear Weapons Act. This measure was supposed to improve the processes behind first use of nuclear weapons and reduce the likelihood of nuclear war by requiring the agreement of declaration of war by Congress – at a time when everyone was concerned that Trump would press the nuclear button just as thoughtlessly as he pressed the tweet button on inflammatory comments to Kim Jong-un. This is the first time that a president's nuclear autonomy has been challenged and reflects concerns about Trump's temperament. The act was shouted down by the Republican majority.

The United Nations (UN) Nuclear Weapon Ban Treaty (Ban Treaty) was introduced in 2017 and has 69 signatories and 19 ratifications at time of writing. Thirty-one more ratifications are required for it to come into force. It is notable that nuclear weapon possessor states and members of NATO declined to vote on this treaty (with the exception of the Netherlands, the only 'No' vote).[3] With their support, nuclear weapons could have become an unnecessary relic of warfare by now, consigned to the past like mustard gas and Zyklon-B. Advances in modern warfare technologies mean that all meaningful military missions can be undertaken using conventional weapons anyway, so the prospect of using nuclear weapons should have become increasingly unthinkable over time.[4] Archaic nuclear weapon systems have already been gazumped by conventional warfare of equivalent scale and impact in conflict scenarios – for instance, the American 'Mother of All Bombs' massive ordnance air blast device (MOAB) was deployed in Afghanistan.[5] However, as Chapter 7 revealed, the norms and taboos surrounding nuclear weapons help to perpetuate their maintenance. They are still seen by nuclear weapon possessor states as irregular and exceptional weapons, rather than the dirty, dangerous, dated old bombs that they are.

However, it is not all doomsday and gloom. There is increasing pressure from nuclear-free zones to support adhesion to the international norms of arms control, non-use, non-proliferation and disarmament. This is not a new phenomenon but a process that has been gradually escalated by non-nuclear weapon possessor states over several decades. Chapter 7 described how the UN Ban Treaty follows on from a progressive trend to limit and abolish nuclear weapons and their specific use since 1945, including normative architecture such as the Non-Proliferation Treaty (NPT), the Comprehensive Test Ban Treaty (CTBT), the Nuclear Suppliers Group, regional NWFZs, the Proliferation Security Initiative and the IAEA.[6]

The Global South is already almost nuclear weapon-free. This is not because they can't afford to join the nuclear club, but because they have chosen a path of positive peace instead. The Tlatelolco Treaty created a nuclear weapon-free Latin America and Caribbean in 1967.[7] In the South Pacific, progress was bloodier. The Treaty of Rarotonga has created a South Pacific NWFZ since August 1985.[8] Sadly, this treaty was accelerated due to a French state-sanctioned terrorist bombing of Greenpeace's

Rainbow Warrior ship. The attack killed a crew member in Auckland, New Zealand on 10 July 1985.[9] This violent intervention by the French occurred because the *Rainbow Warrior* was heading for the Mururoa Atoll in the South Pacific in French Polynesia to disrupt impending French nuclear weapon tests.[10] French agents Captain Dominique Prieur and Commander Alain Mafart[11] were honoured, decorated and promoted once they returned from imprisonment in New Zealand to France, but the South Pacific's nuclear-free future was set.[12]

Other regions with a legacy of harm from nuclear defence activity have followed suit, to create a more peaceful world. Kazakhstan, Kyrgyzstan, Tajikistan, Turkmenistan, and Uzbekistan entered a legally binding commitment during the Semipalatinsk Treaty in 2006, which created the Central Asian NWFZ.[13] These nations are cushioned by Mongolia to the east, which has locked its nuclear-free status in national legislation.[14] Later, the Pelindaba Treaty established the entire continent of Africa as an NWFZ by 2009.[15] These treaties recognise that small-scale inter-state conflicts are not prevented or resolved by nuclear force. By doing so, the Global South has demonstrated that it is more progressive than the North in this dimension. These treaties are significant achievements, contributing towards global security and peace.

The creation of international NWFZs involved disarmament. It is notable that there are more nations that have taken steps towards possessing nuclear weapons and have then subsequently decided to disarm, or been coerced into doing so, than there are states that currently possess nuclear weapons.[16] These nations have recognised nuclear weapons as the expensive liability that they are, rather than a security asset. Some of the states that have chosen to disarm or halt nuclear proliferation are surprising. Taiwan and South Korea halted their plutonium production programmes in the 1960s,[17] Sweden opted for conventional kinetic approaches instead of nuclear weapons, and Argentina and Brazil had both decided against any nuclear aspirations by the 1990s. South Africa dismantled its warheads and abandoned its nuclear programme in 1989,[18] and Ukraine, Kazakhstan and Belarus all decided to surrender the nuclear weapons that they had inherited from the Soviet Union in 1991, in part to support the Central Asian NWFZ.[19] These countries are doing just fine without nuclear weapons, which demonstrates that disarmament is both positive and possible. However, after their nuclear disarmament, Iraq and Libya have suffered long-term fates that provide little incentive for future despots to give up on their nuclear aspirations.[20]

The Ban Treaty is beginning to change the political nuclear culture in Europe. Nuclear weapon possessors are steadily becoming recognised as pariahs, rather than protectors, by nuclear-free states. On the same day that Donald Trump visited Germany in 2018, the Organisation for Security and Co-operation in Europe had 300 parliamentarians from over 53 states in attendance at its 27th Annual Session in Berlin. These parliamentarians expressed their concern about nuclear weapon threats in the region, by proposing a no-first-use policy for nuclear weapons and the adoption of other disarmament and confidence-building measures.[21] While the assembly rejected the removal of all nuclear weapons hosted by non-NATO countries and did not endorse the Ban Treaty, this is a step in the right direction for Europe.

Despite implementing international nuclear weapon testing treaties to protect the uninhabited Antarctic, space and the seabed, the Global North appears to have little regard for the impacts of nuclear weapons on humanity. All the world's oceans are still occupied by nuclear submarines, creating a global risk. The USA, UK, France, China and Russia have maintained a protectionist attitude towards their nuclear weapons. American academics have tried to diminish the peaceful power of nuclear-free zones by describing them as 'understood within a limited framework, and not as a major step towards world disarmament'.[22] A disparity exists between the words and actions of nuclear weapon possessor states such as Russia and the US, where nuclear norms are espoused for other states but not always adhered to by their creators – and it is difficult for international organisations such as the IAEA and UN to ensure that the most powerful states comply.[23] There are still ongoing challenges with Israel, Iran and North Korea. It is notable that Japan, the only state to experience nuclear warfare, has chosen not to endorse the Ban Treaty, despite their geographical vulnerability to the threats in the South Pacific region. The future looks dark, but to understand the possible outcomes for nuclear warfare, critical scenarios, places and activism should be explored.

CRITICAL SCENARIOS

The history, culture and geography of nuclear weapons provide insights into the future of nuclear warfare. Nuclear weapons have been part of the defence portfolio for over 70 years without their use in nuclear warfare, so it is possible that the current status quo could be maintained. Existing conflicts could resurface with states that have attempted to proliferate or already possess nuclear weapons, such as Iran and North Korea. Alternately, voluntary or forced nuclear disarmament could arise if the Ban Treaty is successful. This has serious implications for the power dynamic of the Global North; countries may not want to relinquish their nuclear weapons and could instead opt for a Brexit-style abandonment of the UN. Nuclear security challenges could arise as weapons are dismantled, and there is the perennial challenge of what to do with waste radioactive materials.[24] There is also the potential for unforeseen scenarios to arise, should unanticipated states decide to proliferate, or in the unlikely event that nuclear terrorism occurs.

The uncertainty that surrounds the future of nuclear warfare has meant that nuclear weapon possessor states are resuming civil defence practice and critical scenario studies. It is recognised that existing warning and informing systems and approaches are not fit for purpose in the USA. This was demonstrated by the accidental ballistic missile false alarm in Hawaii on January 2018 that occurred due to a civil defence team mistake. A team of academics in the USA are currently 'Reinventing Civil Defense', by developing new communication strategies for nuclear risk for the twenty-first century.[25] This project notes that while 'discussions of nuclear terrorism, proliferation, and arms control are still regularly in the news ... in many ways it is far easier today to get formal information about nuclear weapons stockpiles, capabilities, and effects than any previous time in history'. Yet something fundamental is missing: the lived experience of nuclear risk has receded for an entire generation of people, worldwide. The Reinventing Civil Defense project aims to use graphic novels, apps, games and virtual reality to provide measured, informed, non-sensationalistic and non-partisan sense of the lived experience of nuclear risk – therefore creating deep awareness and create meaningful public action. This project must avoid the pitfalls of sensationalism or banality that are evident in historic civil defence literature, such as the much-derided UK 'Protect and Survive' programme of the early 1980s.[26] It could have a positive impact on civil defence and nuclear policy if it keeps calm and carries on.

A USA Defense Threat Reduction Agency funded study is also currently under way to try to understand the risk to urban populations should a nuclear attack occur. Geospatial scientists have been crucial to this recent effort and there appears to be a successful academic mutual aid agreement between the USA and the UK. British geographer Andy Crooks and his team at George Mason University are using Geographical Information Systems (GIS), spatial analysis, social network analysis and agent-based modelling methodologies to try and understand individual and social responses to a nuclear WMD event, with focus upon synthesising data and transport networks for New York City, USA.[27] Their work aims to counter and understand the risks of nuclear warfare to the USA, and their burgeoning study has featured prominently in the popular media in the USA and the UK. There is hope that public knowledge of this academic work will allay fears caused by media misinformation about nuclear warfare.

It is not just researchers who are undertaking work to explore critical scenarios relating to a nuclear attack. Another academic geospatial study has recently been completed that explores the outcome of nuclear terrorism and considers the possibility of multiple nuclear detonations to New Delhi, India.[28] In this study, human casualties were simulated to understand the possible spatial distribution of casualties in urban environments, and test a nuclear-specific triage system. This study concludes that improvements need to be made to New Delhi's transport infrastructure to enable mass evacuation, and that existing warning and informing strategies need further development. The nature of this recent research suggests that while nuclear warfare and terrorism are extremely low-likelihood events, they are still perceived as credible threats by nuclear weapon possessor states. The advice of any of these new studies must be carefully considered within the greater context of international geopolitical risk, and it is important that their outcomes do not provide a new route for government scaremongering or political propaganda. Hypothetical scenarios provide scope to theoretically understand risk, but any perception of threat must be grounded in reality.

It is not just Trump and Putin that have increased tensions between the US and Russia since the Cold War. There has been increasing eastward creep by NATO since the Cold War ended, and the US withdrawal from the Anti-Ballistic Missile Treaty, in 2002.[29] The maintenance of negative nuclear peace is dependent upon the future renewal of the NewSTART agreement on 5 February 2021.[30] This agreement limits the USA and Russia to 1,550 deployed strategic nuclear warheads each, and is therefore one of the most important arms control measures in effect.[31] An extension of this agreement beyond 2021 is only possible with both

Russian and American consensus. It must not be forgotten that the US and the Soviet Union were the chief architects of the NPT – a treaty that celebrated its 50th anniversary of opening for signature in 2018. Extending the agreement may be more challenging now than during the previous Medvedev–Obama era, due to the greater volatility and unpredictability of Presidents Putin and Trump.

In the time between the Obama and Trump administrations' nuclear posture reviews, the dynamic between Russia and the USA has become more diffuse and mutable.[32] Despite both states signing the NewSTART Treaty, they have shared but separate intentions to develop their nuclear arsenals. Russia has recently deployed a new intercontinental ballistic missile (ICBM), and has continued or extended the development of three further ICBMs.[33] Meanwhile, Trump plans to implement 'low-yield' mini-nuclear weapons, as an alternative to conventional modes of warfare. The introduction of these low-yield nuclear weapons would erode the nuclear taboo and increase the likelihood of tit-for-tat nuclear warfare. This decision also highlights Trump's lack of interest in leading global efforts to reduce existing nuclear threats and rivalries, unlike his predecessor.

The Helsinki Summit on 16 July 2018 provided more than just a photo op; it also provided a chance for Putin and Trump to plan a future nuclear strategy that involves greater promotion of nuclear security and arms control. However, it is difficult to ascertain the true stability and outcome of these conversations behind closed doors. Following his two-hour conversation with Putin, Trump publicly denounced the American FBI and blamed them for recent Russian cyber-aggressions, then retracted his previous statement a couple of days later.[34] The Republican Speaker of the House Paul Ryan responded to this statement by reminding everyone that Mr Trump 'must appreciate that Russia is not our ally'.

The House Intelligence Committee discovered that Russia had influenced the 2016 US presidential election.[35] Twelve Russian intelligence officers were subsequently charged for hacking the Democratic National Committee by using spear phishing emails and malicious software.[36] This creates further challenges to the US–Russian relationship, in addition to Trump's strangely obsequious behaviour. Putin is the only leader the Trump has not insulted during his time as president, and perhaps there is a strategic element to this on his part, a fear of *kompromat*, or compromising materials from Russian espionage being revealed.[37] Trump's motivations are poorly understood, and it is difficult to ascertain if the threat of nuclear war will increase due to their interactions, or if a volatile and authoritarian United States–Russia pact is on the horizon.

Regardless of the outcomes of future research, relationships and disarmament policies, nuclear terrorism will remain a critical scenario for geographers. Understanding the challenges and risks surrounding fissile material trafficking and the different geospatial scenarios that could occur is crucial for effective preparedness measures. There are still individuals, communities and even states that could support the movement of unpermitted nuclear materials across borders.

CRITICAL PLACES

It is not possible to characterise the exact nature and location of future nuclear weapon disputes or war zones, but volatile regions with the potential for or actual nuclear weapon capacity include Iran, Israel, North Korea, India and Pakistan. Nuclear weapons are an inescapable aspect of twenty-first-century Asia, as the region contains seven states across the spectrum of nuclear weapon status.[38] China was an early member of the 'nuclear club' and an NPT-compliant nuclear weapon possessor state. India and Pakistan both possess nuclear weapons but are not part of the NPT. Unlike the UK and France, India and China have 'no-first-use' policies in place to emphasise the political purpose of their weapons. There are also the umbrella states of Singapore, Japan and South Korea.[39] These states do not possess nuclear weapons, but have bilateral treaty arrangements that have created security alliances with the United States, and play an important role in shaping Asia's nuclear risk.[40] Finally, there is the world's only NPT defector to fulfil its nuclear illicit ambitions, North Korea.

There are still long-lasting troubles with arms control in the Middle East. Iran is a challenging case due to its determination to pursue uranium enrichment under the guise of civilian technology, and its threats to withdraw from the NPT.[41] An Iranian nuclear weapon would cause considerable unrest across the Middle East. This is partially due to the close alliance between the US and Saudi Arabia, a neoliberal arrangement that creates profit for US defence manufacturing and maintains Saudi Arabia's political dominance. Social and material inducements have been applied by the USA and Europe, with limited success, to move Iran away from its burgeoning nuclear weapon programme. Trump declared his intentions to withdraw the nuclear deal with Iran on 8 May 2018. He said: 'This was a horrible one-sided deal that should have never, ever been made.'[42]

This has intensified the toxic relationship that already exists between Iran, America and the Middle East. Trump is politically motivated in abandoning the deal as he attempts to dismantle Obama-era achievements – which he described as 'complicated, technical'.[43] Trump is also supporting Israeli Prime Minister Benjamin Netanyahu and the murderously hard-line Israeli side of Middle East 'peace' negotiations. He declared that 'It doesn't help if I start saying I'm very pro-Israel' when he described his decision to relinquish the Iran deal by citing, in part, evidence presented by Mr Netanyahu.[44] Doubtless, Trump will undertake more interactions of this nature in the future. He has chosen Mike Pompeo and John Bolton as his advisers, men who are described as 'super-hawks' on Iran policy, and could cause further disruption in the Middle East with their influence.[45]

Israel remains a critical nuclear issue. It has been mendacious about its own nuclear stockpile and has recently contested the Iran deal, with Mr Netanyahu claiming that it 'was built on lies'.[46] Perhaps Israel needs more introspection and self-awareness, and to reflect upon its own nuclear history. Every nuclear programme has had murky beginnings, but Israel's is perhaps the most duplicitous. Israel's first Prime Minister David Ben-Gurion quickly declared his nuclear aspirations, soon after Israel was established in 1948. Israel's first deliverable nuclear weapon is thought to have been created in secret collaboration with French scientists by December 1966, further supported by secret shipments of British fissile materials in the 1950s and 1960s.[47] Hypocritically, it developed the Begin Doctrine of counter-proliferation and preventive strikes in the 1960s, denying other regional actors the ability to acquire their own nuclear weapons. Israel has declined to sign the NPT, declaring that it would not be in its security interests. It has neither confirmed nor denied a current arsenal of approximately 80 to 400 warheads.[48] It may also still possess chemical weapons, and appears oblivious to international WMD taboos.[49] Israel's ambiguity leaves it open to nuclear security threats, as it refuses inspection and monitoring under the IAEA. Because of this and its weak strategic location, it may be a source of future nuclear discontent.

North Korea is a volatile authoritarian dictatorship that possesses nuclear weapons in violation of international treaties. It creates new threats that could produce unforeseen and new nuclear geographies. While 'Dear Leader' Kim Jong-un's relationship with President Donald Trump has shown signs of improvement, it is likely that history will reflect upon their interactions since his presidency as representing a time of regression in nuclear politics. However, Trump is not threatening North Korea with WMD at the time of writing. He approached his summit photo op with Kim Jong-un on 12 June 2018 with confidence, declaring: 'I'm very well prepared. I don't think I have to prepare very much. It's about attitude.'[50]

Trump was accompanied by basketball veteran and social media star Dennis Rodman, who is a friend of Kim, and was sponsored by the cannabis cyber-currency 'GreenCoin' to attend the first nuclear summit between President Trump and North Korean leader Kim Jong-un.[51] Rodman and Kim were photographed together laughing like old friends. Trump and Kim's meeting in Singapore aimed to quell nuclear anxieties surrounding North Korean nuclear proliferation, and Trump offered Kim security guarantees in exchange for denuclearisation. Since the summit, North Korea has repatriated the remains of American soldiers who were killed in the Korean War, as a gesture of goodwill. However, North Korea has also undertaken another nuclear weapon test, as of 16 November 2018, in defiance of the US–North Korean Singapore Summit. The outcome of these talks remains unclear. A better understanding of history shows that preparation is needed to ensure successful negotiations in the long term, and that persuasion rather than coercion is crucial.

North Korea is regarded by almost everyone as the greatest future threat to international peace and security. If nuclear war did break out between the USA and North Korea in the future, then the president may have some challenges with Canadian defence policy. There is a Cold War legacy of Canadian concern about the incidental impacts of nuclear weapons that were aimed at the US. For this reason, Canada developed a comprehensive civil defence programme to prepare its citizens for an inadvertent attack.[52] Canada could prove to be an influence on a North Korean attack – the ICBM trajectory would travel over vast swathes of Canadian land – yet if missiles are discovered above Ottawa, then Canada's only military action will be to warn the Americans.[53] This overflight problem has been described as 'exactly the same position that Canada was in during the Cold War' by Fred Armbruster, the executive director of the Canadian Civil Defence Museum.[54]

However, political geography creates huge obstructions to nuclear war, through the nature of existing boundaries, borders and international relationships. It is the revenge of geography that there are even more volatile overflight issues should an American attack on North Korea or China occur – an ICBM launched from Malmstrom Air Force Base in mainland USA would need to penetrate Russian airspace during its trajectory.[55] This would probably be unwise, due to the size of the alert Russian nuclear arsenal and their launch-on-warning doctrine. So, why not launch south, to avoid causing international consternation and possible multilateral nuclear war? A lighter payload and longer timescale would be required, and a fractional orbital bombardment system would be needed to achieve this distance, which was outlawed since 1979 under the Strategic Arms Limitation Talks II, with this retained as a principle by NewSTART I and II.[56] Only nuclear submarines could be effectively used, and from their current base, they would have to travel about 2,500 km from their target to stay out of range of Russia.

Before North Korea developed the bomb, India and Pakistan were the two states of greatest concern, due to the ongoing legacy of their conflict before they attained nuclear arms. India and Pakistan have been destabilised by a history British colonialism, as much as anything else. During the Cold War, India and Pakistan leveraged US–Soviet acrimony into military, economic and diplomatic support from Washington and Moscow.[57] Despite abstention from the NPT and the CTBT, India was subject to sanctions from Australia, Japan and others after undertaking nuclear weapon tests in 1998.[58] India's tests for Operation Shakti were swiftly followed by Pakistani detonation of five nuclear devices in Baluchistan on 28 May 1998.

While the world's attention has been fixed on North Korea, Israel, Iran and the USA; India and Pakistan have been quietly increasing the likelihood of nuclear war. Both states have 'diverted public unrest and revived nationalist sentiments in part through the development of nuclear weapons'.[59] In the 2000s this created a type of negative peace in the region, since only India has a 'no-first-use' policy. However, India recently launched a submarine armed with nuclear weapons. This means that India now has ICBM, bomber and submarine strike capacity, a complete nuclear triad. Before this, Indo-Pakistani proliferation had been land-based. In response, Pakistan has reached a deal with China to buy more diesel-electric attack submarines that can be equipped with nuclear weapons, scheduled for delivery in 2028.[60] Pakistan is also extending its nuclear capacity with both strategic and tactical weapons, and has also sought to develop its second-strike capability through different types of deterrent.[61] This may result in India re-evaluating its strategy of maintaining strategic weapons, in order to dissuade Pakistan from ever using its tactical nuclear arsenal. However, India is currently incentivised to develop a range of similar low-yield tactical nuclear weapons, just in case. 'The nuclearization of the Indian Ocean has begun,' said Zafar Jaspal, a nuclear security expert at Islamabad's Quaid-I-Azam University, 'Both states have now crossed the threshold.'[62] This is an issue, as there is a weakened chain of command and control associated with nuclear submarines, putting India and Pakistan at much greater risk of an accident that could escalate into nuclear war. This is also an issue because of the deadly consequences that could occur if two enemies with a shared border continue to proliferate.

Unlike Pakistan, India has the civilian support of the USA under the Indo-US Nuclear Deal. This agreement separates India's civil and military nuclear facilities, while complying with IAEA safeguards, in exchange for full civilian nuclear cooperation with India.[63] Pakistan came close to using its nuclear weapons as a counter to conventional warfare during the Kargil War in 1999; with international concerns resurfacing again in 2001 during the India–Pakistan standoff. However, when discussing the Ban Treaty, India's foreign ministry spokesman Gopal Baglay said that it 'in no way constitutes or contributes to the development of any customary international law', and that India is not party 'and so shall not be bound by any of the obligations that may arise from it'.[64]

While the theatre of nuclear war is in full-swing, the idea of self-assured destruction (SAD) has been introduced by the group of scientists who pioneered the idea of nuclear winter, and modelled its consequences.[65] It is not just the brute force of nuclear weapons that can kill thousands of people in a geographically limited target zone, as seen in Hiroshima and Nagasaki. There is a more concerning diffuse effect to our entire planet. Nuclear weapon possessor states are so arrogant and nihilistic that they are willing to risk the destruction of humanity to retain their nuclear status, hence 'self-assured destruction'. Despite progress in nuclear disarmament, even a 'small-scale' nuclear conflict between India and Pakistan could induce nuclear winter, by pushing smoke particulates into the global atmosphere, reducing temperatures and destroying the ozone layer that shields us from harmful ultraviolet radiation.[66,67] This would threaten food security and affect every species on the planet.

Pacifist action is ongoing to try to counter the nuclear threat, but the outcome of the UN Ban Treaty is not certain yet. The ideal scenario would be a more peaceful direction for nuclear warfare, where arms control and nuclear security are tightened before international and multilateral disarmament occurs. Even then, it is unlikely that the Global North countries will willingly relinquish their weapons, or that North Korea will comply with international legislation – leaving a dangerous scenario where the most unstable country with nuclear weapons becomes the only country with them. The science and technology that supports nuclear weapon production cannot be un-invented, but with time, the tacit knowledge of the people who design and manufacture nuclear weapons will be lost, practically un-inventing the bomb.[68] There are glimpses of peace. UN Secretary-General Guterres published *Securing our common future* in May 2018, which provides a new agenda for nuclear disarmament.[69] This document provides informed insights into the way that disarmament could be attained, offering hope.

There is also a more mundane aspect of impending risk arising from the UK's nuclear future, as it breaks away from Euratom regulation by the European Atomic Energy Community due to Brexit.[70] This is creating a time of uncertainty, where the operability and effectiveness of current regulations may be put at risk. British politicians have found it hard to vote against nuclear weapons. The UK has remained part of the nuclear club due to a culture of nuclearism, national identity and the local politics of jobs and industry, which are reflected in the vote for Trident renewal in 2016.[71,72] This means that Britain will remain nuclear, but potentially without the overarching continental safeguards that have maintained arms control and nuclear security.

Geographers have had a powerful influence upon warfare by orchestrating zones of conflict and unrest, and supporting geotechnologies for the development of lethal and dirty weapons. From Halford Mackinder's theories of warfare that supported the early Cold War, to Prime Minister Theresa May herself, who studied in the department that Halford McKinder established – the first geographer with both the willingness and capacity to push the nuclear button and kill thousands.

However, geographers are also the gatekeepers of responsible arms control and disarmament. They need to ensure that cartography and other geotechnologies are used to support our understanding of the risks of nuclear weapons and to reduce the likelihood of nuclear war, rather than just representing the ideologies and anxieties of those who commission and create maps. Geographers have the capacity to provide authentic and strong support, and to reveal the many further hidden spatial inequalities and injustices of nuclear warfare that will doubtless emerge in the future. In an age of misinformation, we geographers, anthropologists and social scientists can help to reveal injustices, to explore geopolitics, and hopefully to prevent nuclear war.

Notes

All urls were checked between July and November 2018.

1 The Radical Geography of Nuclear Warfare

1. De Certeau, M., 1985. Practices of space. In *On signs*, edited by M. Blonsky. Johns Hopkins University Press, p. 129.
2. Gray, C. and Sloan, G. (eds), 2000. *Geopolitics, geography and strategy*. Routledge.
3. Alexis-Martin, B. 2019. The geographies of nuclear war. In R. Woodward (ed.) *Future progress in the geographies of warfare*. Routledge.
4. Jones, C.A. and Smith, M.D., 2015. War/law/space: Notes toward a legal geography of war. *Environment and Planning D: Society and Space*, 33(4): 581–591.
5. Alexis-Martin, B. and Davies, T., 2017. Towards nuclear geography: Zones, bodies, and communities. *Geography Compass* 11(9).
6. Brown, K.L., 2013. *Plutopia: Nuclear families, atomic cities, and the great Soviet and American plutonium disasters*. Oxford University Press.
7. Quoted in Amey, M.D., 2005. Living under the bell jar: Surveillance and resistance in Yevgeny Zamyatin's We. *Critical Survey* 17(1): 22–39.
8. Bonnett, A., 2014. *Unruly places: Lost spaces, secret cities, and other inscrutable geographies*. Houghton Mifflin Harcourt.
9. Gentile, M., 2003. Delayed underurbanization and the closed-city effect: The case of Ust'-Kamenogorsk. *Eurasian Geography and Economics* 44(2): 144–156.
10. Bonnett, *Unruly places*.
11. US Department of Energy, Nevada National Security Site leaflet on Mercury: www.nnss.gov/docs/fact_sheets/DOENV_1094.pdf. There is also a PO Box and receiving warehouse if you wish to write to Mercury, but no fixed address: www.nnss.gov/pages/NFO/MOProcurement/MOProcurementInvoice.html
12. Intondi, V., 2018. The dream of Bandung and the UN Treaty on the Prohibition of Nuclear Weapons. *Critical Studies on Security* online first.
13. Gaard, G., 2017. Feminism and environmental justice. In R. Holifield, J. Chakraborty and G. Walker (eds) *The Routledge handbook of environmental justice*. Routledge, pp. 100–114.
14. 'May: Yes I would push the nuclear button': www.bbc.co.uk/news/av/uk-politics-36832530/may-yes-i-would-push-the-nuclear-button
15. Richie, N. 2016. Why Jeremy Corbyn's 'third way' for Trident actually makes sense. 30 January 2016: https://theconversation.com/explainer-why-jeremy-corbyns-third-way-for-trident-actually-makes-sense-53343

16. US Department of State. 1986. Memorandum of conversation. Declassified. 12 October: https://nsarchive2.gwu.edu//NSAEBB/NSAEBB203/Document15.pdf
17. Pepper, D. and Jenkins, A. produced *The geography of peace and war* with Blackwell in 1985.
18. Brown, *Plutopia*.
19. Malin, S.A., 2015. *The price of nuclear power: Uranium communities and environmental justice*. Rutgers University Press.
20. Iversen, K., 2012. *Full body burden: Growing up in the nuclear shadow of Rocky Flats*. Broadway Books.
21. Mason, K., 2017. Ghosts of the future: A normative existentialist critique of nuclear weapons, Mutually Assured Destruction and deterrence. *ACME: An International Journal for Critical Geographies* 16(1): 149–155.
22. See: www.crowdfunder.co.uk/nuclear-refrain

2 A Secret History

1. Gaddis, J.L., 1989. *The long peace: Inquiries into the history of the Cold War*. Oxford University Press.
2. Hoddeson, L., Henriksen, P.W., Meade, R.A. and Westfall, C.L., 1993. *Critical assembly: A technical history of Los Alamos during the Oppenheimer years, 1943–1945*. Cambridge University Press.
3. Kunetka, J.W., 1979. *City of fire: Los Alamos and the atomic age, 1943–1945*. University of New Mexico Press.
4. Badash, L., Hirschfelder, J.O. and Broida, H.P. (eds), 2012. *Reminiscences of Los Alamos 1943–1945*, vol. 5. Springer Science & Business Media.
5. Szasz, F.M., 1992. *British scientists and the Manhattan Project: The Los Alamos years*. Springer.
6. Wellerstein, A., 2016. America at the atomic crossroads. *The New Yorker*: www.newyorker.com/tech/elements/america-at-the-atomic-crossroads
7. Morton, L., 1960. *The decision to use the atomic bomb*, vol. 70, no. 7-23. Center of Military History, US Army.
8. Polmar, N., 2004. *The Enola Gay: The B-29 that dropped the atomic bomb on Hiroshima*. Potomac Books, Inc.
9. Loebs, B., 1995. Hiroshima and Nagasaki. *Commonweal* 122(14): 11.
10. Fedman, D. and Karacas, C., 2012. A cartographic fade to black: Mapping the destruction of urban Japan during World War II. *Journal of Historical Geography* 38(3): 306–328.
11. Cary, O., 1979. Atomic bomb targeting: Myths and realities. *Japan Quarterly* 26(4): 506.
12. Wellerstein, A., 2014. The Kyoto misconception: http://blog.nuclearsecrecy.com/2014/08/08/kyoto-misconception/
13. Wellerstein, A., 2014. The luck of Kokura: http://blog.nuclearsecrecy.com/2014/08/22/luck-kokura/

14. Kristof, N., 1995. Kokura, Japan: Bypassed by A-bomb. *New York Times* 7 August: www.nytimes.com/1995/08/07/world/kokura-japan-bypassed-by-a-bomb.html
15. Malloy, S.L., 2009. Four days in May: Henry L. Stimson and the decision to use the atomic bomb. *Asia-Pacific Journal* 14: 2–9.
16. Lashmar, P., 1994. Stranger than Strangelove: A general's forays into the nuclear zone. *Washington Post*, 3 July: www.washingtonpost.com/archive/opinions/1994/07/03/stranger-than-strangelove-a-generals-forays-into-the-nuclear-zone/4dc9a919-8d2f-47cb-82ee-e74d9ecc436a/?utm_term=.56cbb643dacd
17. Asada, S., 1998. The shock of the atomic bomb and Japan's decision to surrender: A reconsideration. *Pacific Historical Review* 67(4): 477–512.
18. Pape, R.A., 1993. Why Japan surrendered. *International Security* 18(2): 154–201.
19. Stimson, H.L. and Truman, H.S., 1947. The decision to use the atomic bomb. *Bulletin of the Atomic Scientists* 3(2): 37–67.
20. Gar Alperovitz, 1994. *Atomic diplomacy. Hiroshima and Potsdam: The use of the atomic bomb and the American confrontation with Soviet power.* Pluto Press.
21. Aoki, K., 1996. Foreign-ness and Asian American Identities: Yellowface, World War II propaganda, and bifurcated racial stereotypes. *UCLA Asian Pacific American Law Journal* 4: 1.
22. Szasz, F.M., 2009. 'Pamphlets away': The allied propaganda campaign over Japan during the last months of World War II. *Journal of Popular Culture* 42(3): 530–540.
23. Brodie, J.F., 2015. Radiation secrecy and censorship after Hiroshima and Nagasaki. *Journal of Social History* 48(4): 842–864.
24. Keever, B.D., 2004. *News zero: The New York Times and the bomb.* Common Courage Press, p. 16.
25. Hersey, J., 1985. *Hiroshima.* Vintage Books.
26. Burchett, W.G., 1983. *Shadows of Hiroshima.* Verso.
27. Herken, G., 2014. *The winning weapon: The atomic bomb in the Cold War, 1945–1950.* Princeton University Press.
28. Cooney, J.P., 1949. Psychological factors in atomic warfare. *Radiology* 53(1): 104–110.
29. Boyer, P.S., 1998. *Fallout: A historian reflects on America's half-century encounter with nuclear weapons.* Ohio State University Press.
30. DiCarlo, A.L., Maher, C., Hick, J.L., Hanfling, D., Dainiak, N., Chao, N. et al., 2011. Radiation injury after a nuclear detonation: Medical consequences and the need for scarce resources allocation. *Disaster Medicine and Public Health Preparedness* 5(S1): S32–S44.
31. Brodie, Radiation secrecy and censorship after Hiroshima and Nagasaki.
32. Press, D.G., Sagan, S.D. and Valentino, B.A., 2013. Atomic aversion: Experimental evidence on taboos, traditions, and the non-use of nuclear weapons. *American Political Science Review* 107(1): 188–206.

33. Polenberg, R. (ed.), 2002. *In the matter of J. Robert Oppenheimer: The security clearance hearing*. Cornell University Press.
34. Schweber, S.S., 2000. *In the shadow of the bomb: Bethe, Oppenheimer, and the moral responsibility of the scientist*. Princeton University Press.
35. Exhibit 14, 'Fatal Accidents' (ca. late 1946), in Los Alamos Project Y, Book II: Army Organization, Administration, and Operation, copy in Manhattan Project: Official history and documents [microform]. University Publications of America (1977), reel 12.
36. Jacobs, R.A., 2009. *The dragon's tail: Americans face the atomic age*. University of Massachusetts Press.
37. Clark, C., 1997. *Radium girls, women and industrial health reform: 1910–1935*. University of North Carolina Press.
38. Cordle, D., 2017. Sciences/humans/humanities: Dexter Masters' The Accident and Being in the Nuclear Age. *Journal of Literature and Science* 10(2): 74–87.
39. Hoddeson et al., *Critical assembly*.
40. Wellerstein, A., 2016. The demon core and the strange death of Louis Slotin. *New York Times*, 21 May: www.newyorker.com/tech/elements/demoncore-the-strange-death-of-louis-slotin
41. Brode, B., 1980. Tales of Los Alamos. In L. Badash, J.O. Hirschfelder and H.P. Broida (eds) *Reminiscences of Los Alamos 1943–1945*. Springer, pp. 133–159.
42. Howes, R.H. and Herzenberg, C.L., 2003. *Their day in the sun: Women of the Manhattan Project*. Temple University Press.
43. Harvey, J. and Ogilvie, M., 2000. Libby, Leona Woods Marshall. In J. Harvey and M. Ogilvie (eds) *The biographical dictionary of women in science: Pioneering lives from ancient times to the mid-20th century*. Routledge, pp. 787–789.
44. Marshall, J., Herzenberg, C., Howes, R., Weaver, E. and Gans, D., 2010. Women and men of the Manhattan Project. *The Physics Teacher* 48(4): 228–232.
45. Hornig, L.S., 1979. Scientific sexism. *Annals of the New York Academy of Sciences* 323(1): 125–133.
46. Broadhead, L.A., 2009. Our day in their shadow: Critical remembrance, feminist science and the women of the Manhattan Project. *Peace and Conflict Studies* 15(2): 38–61.
47. Johnson, K.E., 1986. Maria Goeppert Mayer: Atoms, molecules and nuclear shells. *Physics Today* 39(9): 44–49.
48. Bugnion, F., 2005. The International Committee of the Red Cross and nuclear weapons: From Hiroshima to the dawn of the 21st century. *International Review of the Red Cross* 87(859): 511–524.
49. McCurry, J. 2016. Story of cities #24: How Hiroshima rose from the ashes of nuclear destruction. *The Guardian*, 18 April: www.theguardian.com/cities/2016/apr/18/story-of-cities-hiroshima-japan-nuclear-destruction
50. Ibid..
51. Taylor, N.A.J. and Jacobs, R., 2015. Re-imagining Hiroshima. *Critical Military Studies* 1: 99–101.

52. Alexis-Martin, B., 2015. The Chernobyl necklace: Psychosocial experiences of female radiation emergency survivors. *BELGEO* 1: 1–10.
53. Neary, I., 2003. Burakumin at the end of history. *Social Research: An International Quarterly* 70(1): 269–294.
54. Yoneyama, L., 1999. *Hiroshima traces: Time, space, and the dialectics of memory*. University of California Press.
55. Rich, M., 2016. Survivors recount horrors of Hiroshima and Nagasaki. *New York Times*, 28 May: www.nytimes.com/2016/05/28/world/asia/survivors-recount-horrors-of-hiroshima-and-nagasaki.html
56. Cho, H., 2011. *Competing futures: War narratives in postwar Japanese architecture 1945–1970*. PhD thesis, University of Southern California.
57. Ito, T., 2015. Reconstruction of Hiroshima industry 1945–1960. 地域経済研究: 広島大学大学院社会科学研究科附属地域経済システム研究センター紀要 26: 3–15.
58. Roberts, K., 2013. Hiroshima Peace Memorial Park: An architectural consignation. In M. Lozanovska (ed.), *Cultural ecology: New approaches to culture, architecture and ecology*. School of Architecture + Built Environment, Deakin University, pp. 66–73.
59. Whalen, P.P., 1982. Status of Los Alamos efforts related to Hiroshima and Nagasaki dose estimates. In V.P. Bond and J.W. Thiessen (eds) *Reevaluations of dosimetric factors: Hiroshima and Nagasaki*. Los Alamos National Laboratory.
60. Tsutomu Yamaguchi's obituary. *The Economist*, 14 January 2010.
61. Alexis-Martin, B., 2017. Life after the bomb: Exploring the psychogeography of Hiroshima. *The Guardian*, 6 August.

3 The Mystery of the X-ray Hands

1. Ashley, J. and Hamilton, A., 1991. Nuclear test veterans. *Hansard*. House of Commons Official Report, 188(80): 375–382.
2. Hacker, B.C., 1992. Radiation safety, the AEC, and nuclear weapons testing. *The Public Historian* 14(1): 31–53.
3. Oulton, W.E., 1987. *Christmas Island cracker: An account of the planning and execution of the British thermo-nuclear bomb tests, 1957*. Thomas Harmsworth Publishing.
4. Alexis-Martin, B., 2016. 'It was a blast!' – Camp life on Christmas Island, 1956–1958. *Environment & Society* 19 (autumn). Rachel Carson Centre for Environment and Society.
5. Pyne, K., 1995. Art or article? The need for and nature of the British hydrogen bomb, 1954–58. *Contemporary British History* 9(3): 562–585.
6. Schwartz, S.I., 2011. *Atomic audit: The costs and consequences of US nuclear weapons since 1940*. Brookings Institution Press.
7. Siracusa, J.M., 2008. *Nuclear weapons: A very short introduction*. Oxford University Press.

8. Titus, A.C., 2001. *Bombs in the backyard: Atomic testing and American politics*. University of Nevada Press.
9. Kunkle, T. and Ristvet, B., 2013. *Castle Bravo – fifty years of legend and lore. A guide to off-site radiation exposures*. DTRIAC-SR-12-001. Defense Threat Reduction Information Analysis Center.
10. Brown, A.L., 2014. No promised land: The shared legacy of the Castle Bravo nuclear test. *Arms Control Today* 44(2): 40.
11. Sagan, S.D., 1997. Why do states build nuclear weapons? Three models in search of a bomb. *International Security* 21(3): 54–86.
12. Khalturin, V.I., Rautian, T.G., Richards, P.G. and Leith, W.S., 2005. A review of nuclear testing by the Soviet Union at Novaya Zemlya, 1955–1990. *Science and Global Security* 13(1–2): 1–42.
13. Tóth, T., 2016. Conflict, cooperation, and the Comprehensive Nuclear-Test-Ban Treaty: Financial markets as a metaphor for cycles in global security. *Nonproliferation Review* 23(3–4): 377–384.
14. Toptayeva, B., 2018. Antinuclear movements in the US and Kazakhstan: A cross-cultural analysis of mass communication patterns. Iowa State University Thesis Repository.
15. Nesipbaeva, K. and Chang, C., *Environmental problems on Kazakhstan and the Nevada-Semipalatinsk antinuclear movement*. IAEA report.
16. Freedman, L., 1980. *Britain and nuclear weapons*. Springer.
17. Gelis, U., 2015. The caretaker and the plague: British nuclear weapons testing in Australia. *Nuclear Age Peace Foundation* 27.
18. Cirincione, J., 2007. *Bomb scare: The history and future of nuclear weapons*. Columbia University Press.
19. Alexis-Martin, B., Waight, E., Blell, M. and Bowler, F., Nuclear Families interview series 2016–2018. University of Southampton.
20. Tharmalingam, S., Sreetharan, S., Kulesza, A.V., Boreham, D.R. and Tai, T.C., 2017. Low-dose ionizing radiation exposure, oxidative stress and epigenetic programming of health and disease. *Radiation Research* 188(4.2): 525–538.
21. Hotez, P.J., 2016. Neglected tropical diseases in the Anthropocene: The cases of Zika, Ebola, and other infections. *PLoS Neglected Tropical Diseases* 10(4): e0004648.
22. Keane, J., 2003. Maralinga's Afterlife. *The Age*, 11 May.
23. Parkinson, A., 2004. The Maralinga rehabilitation project. *Medicine, Conflict and Survival* 20(1): 70–80.
24. Arnold, L. and Pyne, K., 2001. Britain's biggest explosion – Grapple Y. In *Britain and the H-Bomb*. Palgrave Macmillan, pp. 165–175.
25. Berkhouse, L., Davis, S.E., Gladeck, F.R., Hallowell, J.H. and Jones, C.B., 1983. *Operation Dominic I, 1962*. No. KT-82-018 (R). Santa Barbara, CA.
26. Hines, N.O., 1963. *Proving ground: An account of the radiobiological studies in the Pacific, 1946–1961*. University of Washington Press.
27. Alexis-Martin, B., 2016. Grapple slings and moonshine: Conversations with the men who tested atomic weapons on Christmas Island. European Research Council Toxic News.

28. Garcia, B., 1994. Social-psychological dilemmas and coping of atomic veterans. *American Journal of Orthopsychiatry* 64(4): 651–655.
29. Murphy, B.C., Ellis, P. and Greenberg, S., 1990. Atomic veterans and their families: Responses to radiation exposure. *American Journal of Orthopsychiatry* 60(3): 418–427.
30. McLaughlin, R., Nielsen, L. and Waller, M., 2008. An evaluation of the effect of military service on mortality: Quantifying the healthy soldier effect. *Annals of Epidemiology* 18(12): 928–936.
31. Bross, I.D. and Bross, N.S., 1987. Do atomic veterans have excess cancer? New results correcting for the healthy soldier bias. *American Journal of Epidemiology* 126(6): 1042–1050.
32. Darby, S.C., Kendall, G.M., Fell, T.P., O'Hagan, J.A., Muirhead, C.R., Ennis, J.R. et al., 1988. A summary of mortality and incidence of cancer in men from the United Kingdom who participated in the United Kingdom's atmospheric nuclear weapon tests and experimental programmes. *British Medical Journal (Clinical Research Ed.)* 296(6618): 332–338.
33. Muirhead, C.R., Bingham, D., Haylock, R.G.E., O'Hagan, J. A., Goodill, A.A., Berridge, G.L.C. et al., 2003. Follow up of mortality and incidence of cancer 1952–98 in men from the UK who participated in the UK's atmospheric nuclear weapon tests and experimental programmes. *Occupational and Environmental Medicine* 60(3): 165–172.
34. Muirhead, C.R., Kendall, G.M., Darby, S.C., Doll, R., Haylock, R.G.E., O'Hagan, J.A. et al., 2004. Epidemiological studies of UK test veterans, II: Mortality and cancer incidence. *Journal of Radiological Protection* 24(3): 219.
35. Roff, S.R., 1999. Mortality and morbidity of members of the British Nuclear Tests Veterans Association and the New Zealand Nuclear Tests Veterans Association and their families. *Medicine, Conflict, and Survival* 15: i–ix.
36. Roff, S.R., 1998. Puff the magic dragon: How our understanding of fallout, residual and induced radiation evolved over fifty years of nuclear weapons testing. *Medicine, Conflict and Survival* 14(2): 106–119.
37. Gun, R.T., Parsons, J., Crouch, P., Ryan, P. and Hiller, J.E., 2008. Mortality and cancer incidence of Australian participants in the British nuclear tests in Australia. *Occupational and Environmental Medicine*.
38. Rivas, M., Rojas, E., Araya, M.C. and Calaf, G.M., 2015. Ultraviolet light exposure, skin cancer risk and vitamin D production. *Oncology Letters* 10(4): 2259–2264.
39. Menvielle, G., Fayossé, A., Radoï, L., Guida, F., Sanchez, M., Carton, M. et al., 2016. The joint effect of asbestos exposure, tobacco smoking and alcohol drinking on laryngeal cancer risk: Evidence from the French population-based case-control study, ICARE. *Occupational and Environmental Medicine* 73(1): 28–33.
40. Boice, J.D., 2017. *Epidemiologic study of one million US radiation workers and veterans*. National Council on Radiation Protection and Measurements No. DOE-NCRP-0008944. Bethesda, MD.

41. Trundle, C., 2011. Biopolitical endpoints: Diagnosing a deserving British nuclear test veteran. *Social Science & Medicine* 73(6): 882–888.
42. Radiation Exposure Compensation Act, www.ncbi.nlm.nih.gov/books/NBK221729/
43. Schwartz, *Atomic audit*.
44. Hansen, D. and Schriner, C., 2005). Unanswered questions: The legacy of atomic veterans. *Health Physics* 89(2): 155–163.
45. Zaretsky, N., 2015. Radiation suffering and patriotic body politics in the 1970s and 1980s. *Journal of Social History* 48(3): 487–510.
46. Rabbitt Roff, S.U.E., 2004. Establishing the possible radiogenicity of morbidity and mortality from participation in UK nuclear weapons development. *Medicine, Conflict and Survival* 20(3): 218–241.
47. Roff, Mortality and morbidity of members of the British Nuclear Tests Veterans Association.
48. Alexis-Martin, B. 2016. Keep the home fires burning: Rekindling the flame with the French and British nuclear test veterans. University of Southampton Faculty of Social, Human & Mathematical Sciences, 6 July.

4 After Nuclear Imperialism

1. UNSCEAR, 2008. *Report of the United Nations Scientific Committee on the Effects of Atomic Radiation.* UN Scientific Committee on the Effects of Atomic Radiation: http://www.unscear.org/unscear/en/publications/2008_1.html
2. Danielssons, P.R., 1986. *French Nuclear colonialism in the Pacific*. Penguin Australia.
3. Kahn, M., 2000. Tahiti intertwined: Ancestral land, tourist postcard, and nuclear test site. *American Anthropologist* 102(1): 7–26.
4. Jorgenson, T. 2016. Bikini islanders still deal with fallout of US nuclear tests, 70 years later. *The Conversation*, 29 June: https://theconversation.com/bikini-islanders-still-deal-with-fallout-of-us-nuclear-tests-70-years-later-58567
5. Maralinga Tours website title page: Maralinga Tours – discover Australia's shocking atomic legacy: https://maralingatours.com.au
6. Kuletz, V., 2001. Invisible spaces, violent places: Cold War nuclear and militarized landscapes. *Violent Environments*, proceedings of a workshop. Cornell University Press, pp. 237–260.
7. Power, P.F., 1986. The South Pacific nuclear-weapon-free zone. *Pacific Affairs* 59(3): 455–475.
8. Kuletz, V.L., 2016. *The tainted desert: Environmental and social ruin in the American West*. Routledge.
9. Ruff, T.A., 2015. The humanitarian impact and implications of nuclear test explosions in the Pacific region. *International Review of the Red Cross* 97(899): 775–813.
10. Ghettas, M.L., 2017. *Algeria and the Cold War: International Relations and the Struggle for Autonomy*. IB Tauris.

11. Keown, M., Taylor, A. and Treagus, M. (eds), 2018. *Anglo-American imperialism and the Pacific: Discourses of encounter.* Routledge.
12. Threet, J.B., 2005. Testing the bomb: Disparate impacts on Indigenous Peoples in the American West, the Marshall Islands, and in Kazakhstan. *University of Baltimore Journal of Environmental Law* 13: 29.
13. Chitkara, M.G., 1996. *Toxic Tibet under nuclear China.* APH Publishing.
14. Arnold, L. and Smith, M., 2006. *Britain, Australia and the bomb: The nuclear tests and their aftermath.* Springer.
15. Reed, T.C. and Stillman, D.B., 2010. *The nuclear express: A political history of the bomb and its proliferation.* Zenith Press.
16. Walsh, J., 1997. Surprise down under: The secret history of Australia's nuclear ambitions. *The Nonproliferation Review* 5(1): 1–20.
17. Taylor, J., 2011. Postcolonial transformation of the Australian Indigenous population. *Geographical Research* 49(3): 286–300.
18. Goodall, H., 1992. 'The whole truth and nothing but ...': Some intersections of western law, Aboriginal history and community memory. *Journal of Australian Studies* 16(35): 104–119.
19. Brady, M., 2017. Atomic thunder: The Maralinga story. Book review. *Aboriginal History* 41: 235.
20. Mattingley, C. and Hampton, K., 1988. *Survival in our own land: 'Aboriginal' experiences in 'South Australia' since 1836.* Wakefield Press Pty, Ltd.
21. Gilbert, H., 2013. Indigeneity, time and the cosmopolitics of postcolonial belonging in the Atomic Age. *Interventions* 15(2): 195–210.
22. Arnold and Smith, *Britain, Australia and the bomb.*
23. Eames, G., 1985. *Final submission by counsel on behalf of Aboriginal organisations and individuals [to the] Royal Commission into British nuclear tests in Australia*: https://trove.nla.gov.au/work/35376237?q&versionId=43993258
24. Wise, K.N. and Moroney, J.R., 1992. *Public health impact of fallout from British nuclear weapons tests in Australia, 1952–1957.* No. ARL-TR-105. Australian Radiation Lab.
25. Williams, G.A., O'Brien, R.S., Grzechnik, M. and Wise, K.N., 2017. Estimates of radiation doses to the skin for people camped at Wallatinna during the UK Totem 1 atomic weapons test. *Radiation Protection Dosimetry* 174(3): 322–336.
26. Mittmann, J.D., 2017. Maralinga: Aboriginal poison country. *Agora* 52(3): 25.
27. BBC, 2014. Lingering impact of British tests on the Australian outback. BBC News, 31 December: www.bbc.co.uk/news/world-australia-30640338
28. Frankel, M., Scouras, J. and Ullrich, G., 2015. *The uncertain consequences of nuclear weapons use.* Johns Hopkins University, Laurel MD, Applied Physics Lab.
29. In Michel, D., 2003. Villains, victims and heroes: Contested memory and the British nuclear tests in Australia. *Journal of Australian Studies* 27(80): 221–228.
30. Keane, J., 2003. Maralinga's afterlife. *The Age*, 11 May.

31. Cross, R., 2004. British nuclear tests and the Indigenous people of Australia. In Barnaby, F. and Holdstock, D. (eds), *The British nuclear weapons programme, 1952–2002*. Routledge, pp. 93–106.
32. Keane, Maralinga's afterlife.
33. Lokan, K.H., 1985. *Residual radioactive contamination at Maralinga and Emu, 1985*. No. ARL/TR--070. Australian Radiation Lab.
34. Parkinson, A., 2002. Articles on the Maralinga 'clean-up' by nuclear engineer and whistle-blower Alan Parkinson. *Journal of the International Physicians for the Prevention of Nuclear War* 7(2): 77–81.
35. ABC News, 2014. Maralinga: Traditional owners given unrestricted access to former nuclear testing site. 5 November: https://mobile.abc.net.au/news/2014-11-05/traditional-owners-given-maralinga-unrestricted-access/5865048?pfm=sm&topic=latest
36. Ibid.
37. Borg, M., 2017. Little-known South Australian history: Uncovering the truth behind the nuclear weapons project at Maralinga. *Oral History Australia Journal* 39: 23.
38. Beck, H.L. and Bennett, B.G., 2002. Historical overview of atmospheric nuclear weapons testing and estimates of fallout in the continental United States. *Health Physics* 82(5): 591–608.
39. Hirshberg, L., 2012. Nuclear families: (Re-)producing 1950s suburban America in the Marshall Islands. *Organization of American Historians Magazine of History* 26(4): 39–43.
40. Ruff, The humanitarian impact and implications of nuclear test explosions in the Pacific region.
41. Parsons, K.M. and Zaballa, R.A., 2017. *Bombing the Marshall Islands: A Cold War tragedy*. Cambridge University Press.
42. Hirshberg, Nuclear families.
43. Ibid.
44. Black-Branch, J.L. and Fleck, D. (eds), 2016. *Nuclear non-proliferation in international law*, vol. III: *Legal aspects of the use of nuclear energy for peaceful purposes*. Springer.
45. Weisgall, J.M., 1980. The nuclear nomads of Bikini. *Foreign Policy* 39: 74–98.
46. Connell, J., 2012. Population resettlement in the Pacific: Lessons from a hazardous history? *Australian Geographer* 43(2): 127–142.
47. Conard, R.A., 1984. Late radiation effects in Marshall Islanders exposed to fallout 28 years ago. In J.D. Boice (ed.) *Radiation carcinogenesis: epidemiology and biological significance*. Raven Press.
48. Gerrard, M.B., 2015. America's forgotten nuclear waste dump in the Pacific. *SAIS Review of International Affairs* 35(1): 87–97.
49. *The Guardian*. 2016. Marshall Islands nuclear arms lawsuit thrown out by UN's top court. 6 October: www.theguardian.com/world/2016/oct/06/marshall-islands-nuclear-arms-lawsuit-thrown-out-by-uns-top-court
50. Maclellan, N., Deery, P. and Kimber, J., 2015. Grappling with the bomb: Opposition to Pacific nuclear testing in the 1950s. In *Proceedings of the 14th*

Biennial Labour History Conference, February. Australian Society for the Study of Labour History, p. 21.
51. Bolton, M. 2018. The devastating legacy of British and American nuclear testing at Kiritimati (Christmas) and Malden Islands. Just Security: www.justsecurity.org/56127/devastating-legacy-british-american-nuclear-testing-kiritimati-christmas-malden-islands/
52. Maclellan, N., 2005. The nuclear age in the Pacific islands. *The Contemporary Pacific* 17(2): 363–372.
53. Ibid.
54. Maclellan et al., Grappling with the bomb.
55. Prăvălie, R., 2014. Nuclear weapons tests and environmental consequences: A global perspective. *Ambio* 43(6): 729–744.
56. Bolton, M., 2017. Humanitarian and environmental action to address nuclear harm. Background paper. International Disarmament Institute, Pace University.
57. Johnston, B.R., 2010. Social responsibility and the anthropological citizen. *Current Anthropology* 51(S2): S235–S247.
58. Agence France Presse. 2014. Marshall Islands want US to resolve unfinished nuclear legacy. 1 March.
59. Bergkvist, N.O. and Ferm, R., 2000. *Nuclear explosions 1945–1998*. No. FOA-R--00-01572-180. Defence Research Establishment.
60. Khalturin, V.I., Rautian, T.G., Richards, P.G. and Leith, W.S., 2005. A review of nuclear testing by the Soviet Union at Novaya Zemlya, 1955–1990. *Science and Global Security*, 13(1–2): 1–42.
61. Norris, R.S. and Arkin, W.M., 1998. Soviet nuclear testing, August 29, 1949–October 24, 1990. *Bulletin of the Atomic Scientists* 54(3): 69.
62. Kassenova, T., 2016. Banning nuclear testing: Lessons from the Semipalatinsk nuclear testing site. *The Nonproliferation Review* 23(3–4): 329–344.
63. Stawkowski, M.E., 2016. 'I am a radioactive mutant': Emergent biological subjectivities at Kazakhstan's Semipalatinsk Nuclear Test Site. *American Ethnologist* 43(1): 144–157.
64. Bauer, S., Gusev, B.I., Pivina, L.M., Apsalikov, K.N. and Grosche, B., 2005) Radiation exposure due to local fallout from Soviet atmospheric nuclear weapons testing in Kazakhstan: Solid cancer mortality in the Semipalatinsk historical cohort, 1960–1999. *Radiation Research* 164(4): 409–419.
65. Ibid.
66. Wegmarshaus, G.R., 2018. Nuclear power and civil society in post-Soviet Russia. In D.R. Marples and M.J. Young (eds) *Nuclear energy and security in the former Soviet Union*. Routledge, pp. 99–119.
67. Bauer, S., 2018. Radiation science after the Cold War: The politics of measurement, risk, and compensation in Kazakhstan. In O. Zvonareva E. Popova and K. Horstman (eds) *Health, technologies, and politics in post-Soviet settings*. Palgrave Macmillan, pp. 225–249.
68. Brunn, S.D., 2011. Fifty years of Soviet nuclear testing in Semipalatinsk, Kazakhstan: Juxtaposed worlds of blasts and silences, security and risks,

denials and memory. In *Engineering Earth: The impacts of mega-engineering projects*. Springer, pp. 1789–1818.
69. Kawano, N., Hirabayashi, K., Matsuo, M., Taooka, Y., Hiraoka, T., Apsalikov, K.N. et al., 2006. Human suffering effects of nuclear tests at Semipalatinsk, Kazakhstan: Established on the basis of questionnaire surveys. *Journal of Radiation Research* 47(Suppl. A): A209–A217.
70. Purvis-Roberts, K.L., Werner, C.A. and Frank, I., 2007. Perceived risks from radiation and nuclear testing near Semipalatinsk, Kazakhstan: A comparison between physicians, scientists, and the public. *Risk Analysis: An International Journal* 27(2): 291–302.
71. Gupta, V., 1995. Locating nuclear explosions at the Chinese test site near Lop Nor. *Science & Global Security* 5(2): 205–244.
72. Bhattacharya, A., 2003. Conceptualising Uyghur separatism in Chinese nationalism. *Strategic Analysis* 27(3): 357–381.
73. Shichor, Y., 2015. See no evil, hear no evil, speak no evil: Middle Eastern reactions to rising China's Uyghur crackdown. *Griffith Asia Quarterly* 3(1): 62–85.
74. *Japan Times*. 2012. China nuclear tests prompt Uighur campaign. 9 August: www.japantimes.co.jp/news/2012/08/09/national/china-nuclear-tests-prompt-uighur-campaign/#.W9h92Hv7Sp0
75. Roberts, B., Manning, R.A. and Montaperto, R.N., 2000. China: The forgotten nuclear power. *Foreign Affairs* 79: 53.
76. Roberts, S., 2012. Imaginary terrorism? The global war on terror and the narrative of the Uyghur terrorist threat. Institute for European, Russian and Eurasian Studies (IERES), Elliott School of International Affairs.

5 After Nuclear War

1. Gusterson, H., 2004. Nuclear tourism. *Journal for Cultural Research* 8(1): 23–31.
2. Kirsch, S., 1997. Watching the bombs go off: Photography, nuclear landscapes, and spectator democracy. *Antipode* 29(3): 227–255.
3. Laurie, I.C., 1979. *Nature in cities*. Wiley.
4. Bennett, L., 2011. The bunker: Metaphor, materiality and management. *Culture and Organization* 17(2): 155–173.
5. Ibid.
6. Schofield, J. and Cocroft, W. (eds), 2009. *A fearsome heritage: Diverse legacies of the Cold War*. Left Coast Press.
7. Woodward, R., 2014. Military landscapes: Agendas and approaches for future research. *Progress in Human Geography* 38(1): 40–61.
8. Swart, J.A.A., 2017. Restoring layered landscapes: History, ecology, and culture. *Environmental Ethics* 39(1): 109–112.
9. Banerjee, S.B., 2000. Whose land is it anyway? National interest, indigenous stakeholders, and colonial discourses: The case of the Jabiluka uranium mine. *Organization & Environment* 13(1): 3–38.

10. Maret, S., 2018. Review, Doom towns: The people and landscapes of atomic testing, a graphic history. *Secrecy and Society* 1(2): 11.
11. Alexis-Martin, B. and Davies, T., 2017. Towards nuclear geography: Zones, bodies, and communities. *Geography Compass* 11(9): e12.
12. Hourdequin, M. and Havlick, D.G. (eds), 2016. *Restoring layered landscapes: History, ecology, and culture.* Oxford University Press.
13. Endres, D., 2018. The most nuclear-bombed place: Ecological implications of the US nuclear testing program. In B. McGreavy, J. Wells, G.F. McHendry Jr and S. Senda-Cook (eds), *Tracing rhetoric and material life.* Palgrave Macmillan, pp. 253–287.
14. Ibid.
15. Alexis-Martin, B. and Malin, S., 2017. An unnatural history of Rocky Flats National Wildlife Refuge, Colorado. Environment & Society Portal, *Arcadia* 25. Rachel Carson Centre for Environment and Society.
16. Hartmann, R., 2014. Dark tourism, thanatourism, and dissonance in heritage tourism management: New directions in contemporary tourism research. *Journal of Heritage Tourism* 9(2): 166–182.
17. Carvalho, B. and Carvalho, P., 2017. Nuclear tourism: From tragedy to adventure. *TURyDES: Revista Turismo y Desarrollo Local* 10(23).
18. Rush-Cooper, N., 2013. *Exposures: Exploring selves and landscapes in the Chernobyl Exclusion Zone* (Doctoral dissertation, Durham University).
19. Woodward, Military landscapes.
20. Ibid.
21. Alexis-Martin, B., 2019. *Geographies of nuclear warfare: A research agenda for military geography.* Edward Elgar Press.
22. Lowe, D., Atherton, C. and Miller, A (eds), 2018. *The unfinished atomic bomb: Shadows and reflections.* Lexington Books.
23. Campbell, D., 1982. *War plan UK: The truth about civil defence in Britain.* Vintage.
24. Vale, L.J., 1987. *The limits of civil defence in the USA, Switzerland, Britain and the Soviet Union: The evolution of policies since 1945.* Springer.
25. Davis, T.C., 2007. *Stages of emergency: Cold War nuclear civil defense.* Duke University Press.
26. Lashmar, P. 2015. Stranger than Strangelove: How the US planned for nuclear war in the 1950s. *The Conversation*, 28 December: https://theconversation.com/stranger-than-strangelove-how-the-us-planned-for-nuclear-war-in-the-1950s-52626
27. Ibid.
28. Conversations with Luke Bennett, Sheffield Hallam. Unpublished archival work. Thank you! 2018.
29. Manchester City Council. 2018. Nuclear Free Authorities: https://secure.manchester.gov.uk/info/500002/council_policies_and_strategies/1130/nuclear_free_local_authorities/1
30. Heuser, B., Heier, T. and Lasconjarias, G., 2018. Military exercises: Political messaging and strategic impact. NATO Forum Paper 26. NATO Defense College.

31. Bennett, L., 2017. *In the ruins of the Cold War bunker: Affect, materiality and meaning-making*. Rowman & Littlefield International.
32. Shapiro, M. and Bird-David, N., 2017. Routinergency: Domestic securitization in contemporary Israel. *Environment and Planning D: Society and Space* 35(4): 637–655.
33. Perry, R.W., Lindell, M.K. and Tierney, K.J. (eds), 2001. *Facing the unexpected: Disaster preparedness and response in the United States*. Joseph Henry Press.
34. Lee'Ann, D., 2013. *The preppers: A multiple case study of individuals who choose a moderate survivalist lifestyle*. PhD dissertation, Northcentral University, Minnesota.
35. Luke Bennett too!
36. Kabel, A. and Chmidling, C., 2014. Disaster prepper: Health, identity, and American survivalist culture. *Human Organization* 73(3): 258–266.
37. Atkinson, R., 2016. Limited exposure: Social concealment, mobility and engagement with public space by the super-rich in London. *Environment and Planning A* 48(7): 1302–1317.
38. Bronson, P., 2013. *The nudist on the lateshift: and other tales of Silicon Valley*. Random House.
39. Alexis-Martin, B., 2016. Colorado's survivalists hunker down for the election – and the apocalypse. *The Conversation*, 31 October: https://theconversation.com/colorados-survivalists-hunker-down-for-the-election-and-the-apocalypse-67475
40. Foster, G.A., 2016. Consuming the apocalypse: Marketing bunker materiality. *Quarterly Review of Film and Video* 33(4): 285–302.
41. Ibid.

6 Strange Cartographies and War Games

1. Freeman, L., 2015. Magic geography of the Cold War. The New School for Social Research. www.publicseminar.org/2015/12/on-inexactitude-in-science/
2. Ibid.
3. Thakur, R., 2018. Japan and the Nuclear Weapons Prohibition Treaty: The wrong side of history, geography, legality, morality, and humanity. *Journal for Peace and Nuclear Disarmament* 1(1): 11–31.
4. Grosser, P., 2016. Timothy Barney, Mapping the Cold War: Cartography and the framing of America's international power. *ERIS – European Review of International Studies* 3(2).
5. Foster, G.A., 2014. Disposable bodies. In G.A. Foster (ed.), *Hoarders, doomsday preppers, and the culture of apocalypse*. Palgrave Pivot, pp. 1–19.
6. Wisner, B., 1986. Geography: War or peace studies? *Antipode* 18(2): 212–217.
7. Alexis-Martin, B. and Davies, T., 2017. Towards nuclear geography: Zones, bodies and communities. *Geography Compass* 11(9): 1–13.
8. Kirsch, S., 2000. Peaceful nuclear explosions and the geography of scientific authority. *The Professional Geographer* 52(2): 179–192.

9. Speier, H., 1941. Magic geography. *Social Research* 8(3):.310–330.
10. Wisner, Geography: War or peace studies?
11. Cutter, S.L., Holcomb, H.B. and Shatin, D., 1986. Spatial patterns of support for a nuclear weapons freeze. *The Professional Geographer* 38(1): 42–52.
12. Cutter, S.L., 1993. *Living with risk: The geography of technological hazards*. Edward Arnold.
13. Openshaw, S., Carver, S. and Fernie, J., 1989. *Britain's nuclear waste: Siting and safety*. Belhaven Press.
14. Openshaw, S., Forbes, G., Miller, E. and Schmalz, R., 1992. The safety and siting of nuclear power plants when faced with terrorism and sabotage. In S.K. Majumdar (ed.), *Natural and technological disasters: Causes, effects and preventive measures*. Pennsylvania Academy of Science, pp. 455–468.
15. Openshaw, S. and Steadman, P., 1982. On the geography of a worst case nuclear attack on the population of Britain. *Political Geography Quarterly* 1(3): 263–278.
16. Bunge, W., 1989. *Nuclear war atlas*. Blackwell.
17. Freeman, Magic geography of the Cold War.
18. Ibid.
19. Barnes, T.J., 2017. A marginal man and his central contributions: The creative spaces of William ('Wild Bill') Bunge and American geography. *Environment and Planning A: Economy and Space* 50(8): 1697–1715.
20. Fedman, D. and Karacas, C., 2012. A cartographic fade to black: Mapping the destruction of urban Japan during World War II. *Journal of Historical Geography* 38(3): 306–328.
21. MacDonald, F., 2007. Anti-Astropolitik – outer space and the orbit of geography. *Progress in Human Geography* 31(5): 592–615.
22. Corson, M.W. and Palka, E.J., 2004. Geotechnology, the US military, and war. In S.D. Brunn, S.L. Cutter and J.W. Harrington (eds), *Geography and technology*. Springer, pp. 401–427.
23. Dalton, C.M., 2013. Sovereigns, spooks, and hackers: An early history of Google geo services and map mashups. *Cartographica: The International Journal for Geographic Information and Geovisualization* 48(4): 261–274.
24. Anderson, B. and Adey, P., 2011. Affect and security: Exercising emergency in 'UK civil contingencies'. *Environment and Planning D: Society and Space* 29(6): 1092–1109.
25. Dittmer, J., 2015. Playing geopolitics: Utopian simulations and subversions of international relations. *GeoJournal* 80(6): 909–923.
26. Miller, C.R., 2003. Writing in a culture of simulation. In M. Nystrand and J. Duffy (eds), *Towards a rhetoric of everyday life: new directions in research on writing, text, and discourse*, University of Wisconsin Press, p. 58.
27. Edwards, P.N., 1997. *The closed world: Computers and the politics of discourse in Cold War America*. MIT Press.
28. Power, M., 2007. Digitized virtuosity: Video war games and post-9/11 cyber-deterrence. *Security Dialogue* 38(2): 271–288.

29. Davis, P.K., Bankes, S.C. and Kahan, J.P., 1986. *A new methodology for modeling national command level decisionmaking in war games and simulations*. No. RAND/R-3290-NA. RAND Corp.
30. Giddens, A., 1994. Living in a post-traditional society. In U. Beck, A. Giddens and S. Lash (eds), *Reflexive modernities: Politics, tradition and aesthetics in the modern social order*. Stanford University Press.
31. Power, Digitized virtuosity.
32. Foster, G.A., 2016. Consuming the apocalypse, marketing bunker materiality. *Quarterly Review of Film and Video* 33(4): 285–302.
33. Brown, K.L., 2013. *Plutopia: Nuclear families, atomic cities, and the great Soviet and American plutonium disasters*. Oxford University Press.
34. Fraser, E., 2016. Awakening in ruins: The virtual spectacle of the end of the city in video games. *Journal of Gaming & Virtual Worlds* 8(2): 177–196.
35. Vanderbilt, T., 2002. *Survival city: Adventures among the ruins of atomic America*. Princeton Architectural Press.
36. *Fallout* Wiki.
37. Masco, J., 2009. Life underground: Building the bunker society. *Anthropology Now* 1(2): 13–29.
38. Alexis-Martin, B., 2018. Interviews with apocalyptic gamers. University of Southampton, UK.
39. Ibid.
40. Parker, M.S., 2009. *Postmodernism and Cold War military technology in the fiction of Don DeLillo and William S. Burroughs*. PhD dissertation, University of Sheffield.
41. Fraser, E., 2016. Awakening in ruins: The virtual spectacle of the end of the city in video games. *Journal of Gaming & Virtual Worlds* 8(2): 177–196.
42. Klinke, I., 2016. Self-annihilation, nuclear play and West Germany's compulsion to repeat. *Transactions of the Institute of British Geographers* 41(2): 109–120.

7 Spaces of Irregularity

1. Sagan, S.D., 1997. Why do states build nuclear weapons? Three models in search of a bomb. *International security* 21(3): 54–86.
2. Tannenwald, N., 1999. The nuclear taboo: The United States and the normative basis of nuclear non-use. *International Organization* 53(3): 433–468.
3. Thakur, R., 2018. Japan and the Nuclear Weapons Prohibition Treaty: The wrong side of history, geography, legality, morality, and humanity. *Journal for Peace and Nuclear Disarmament* 1(1): 11–31.
4. Morgenthau, H.J., 2018. The fallacy of thinking conventionally about nuclear weapons. In P. Foradori, G. Giacomello and A. Pascolini (eds), *Arms control and disarmament*. Palgrave Macmillan, pp. 79–89.
5. Norris, R.S. and Kristensen, H.M., 2011. US tactical nuclear weapons in Europe, 2011. *Bulletin of the Atomic Scientists* 67(1): 64–73.

6. Sanders, D., 2001. Strategic nuclear weapons. In D. Sanders (ed.), *Security co-operation between Russia and Ukraine in the post-Soviet Era*. Palgrave Macmillan, pp. 53–97.
7. Lifton, R.J. and Falk, R., 1982. *Indefensible weapons: The political and psychological case against nuclearism*. Basic Books.
8. Richie, N. 2016. Why British politicians find it so hard to vote against nuclear weapons. *The Conversation*, 19 July: https://theconversation.com/why-british-politicians-find-it-so-hard-to-vote-against-nuclear-weapons-62655
9. Hanson, M., 2018. Normalizing zero nuclear weapons: The humanitarian road to the Prohibition Treaty. *Contemporary Security Policy* 39(3): 464–486.
10. Ibid.
11. Rublee, M.R., 2009. *Nonproliferation norms: Why states choose nuclear restraint*. University of Georgia Press.
12. Alexis-Martin, B., Malin, S., Iversen, K., Sullivan, K. and Blell, M., 2017. In the shadow of Fat Man and Little Boy: How the stigma of nuclear war was unravelled. *The Guardian*: www.theguardian.com/science/brain-flapping/2017/sep/15/in-the-shadow-of-fat-man-and-little-boy-how-the-stigma-of-nuclear-war-was-unravelled
13. Sokolski, H.D., 2004. *Getting MAD: Nuclear mutual assured destruction, its origins and practice*. DIANE Publishing.
14. Jervis, R., 2002. Mutual assured destruction. *Foreign Policy* 133: 40.
15. Reagan, R., 1984. Address before a Joint Session of the Congress on the State of the Union. 24 January: www.presidency.ucsb.edu/ws/?pid=40205
16. Tannenwald, N., 2005. Stigmatizing the bomb: Origins of the nuclear taboo. *International Security* 29(4): 5–49.
17. Ehrhart, H.G., 2017. Postmodern warfare and the blurred boundaries between war and peace. *Defense & Security Analysis* 33(3): 263–275.
18. Mann, M., 1997. Has globalization ended the rise and rise of the nation-state? *Review of International Political Economy* 4(3): S. 472–496.
19. Mattox, J.M., 2010. Nuclear terrorism: The 'other' extreme of irregular warfare. *Journal of Military Ethics* 9(2): 160–176.
20. Sundaram, K. and Ramana, M.V., 2018. India and the policy of no first use of nuclear weapons. *Journal for Peace and Nuclear Disarmament* 1(1): 152–168.
21. Mattox, Nuclear terrorism.
22. Ehrhart, Postmodern warfare and the blurred boundaries between war and peace.
23. Jones, C.A. and Smith, M.D., 2015. War/law/space: Notes toward a legal geography of war. *Environment and Planning D* 33: 581–591.
24. Sokolski, *Getting MAD*.
25. Freedman, L., 2013. Disarmament and other nuclear norms. *The Washington Quarterly* 36(2): 93–108.
26. Collina, T. et al., 2010. The case for the New Strategic Arms Control Treaty. Arms Control Association: www.armscontrol.org/node/4610
27. BBC News, 2015. Nobel Secretary regrets Obama's peace prize. BBC, 17 September: www.bbc.co.uk/news/world-europe-34277960

28. Long, A., 2018. *Russian nuclear forces and prospects for arms control*. RAND Corporation.
29. Petroni, G., 2018. Will Trump and Putin discuss nuclear stability? *Medill News*, 12 July: http://dc.medill.northwestern.edu/blog/2018/07/12/will-trump-putin-discuss-nuclear-stability/#sthash.qYiKnnLd.dpbs
30. Center for Defense Information, 1981. U.S. nuclear weapons accidents: Danger in our midst. *The Defense Monitor* X(5): http://docs.nrdc.org/nuclear/files/nuc_81010001a_n22.pdf
31. Varble, D., 2008. *The Suez crisis*. Rosen Publishing Group.
32. Haftendorn, H., 2011. NATO and the Arctic: Is the Atlantic alliance a Cold War relic in a peaceful region now faced with non-military challenges? *European Security* 20(3): 337–361.
33. Union of Concerned Scientists, 2015. Close calls with nuclear weapons: www.ucsusa.org/sites/default/files/attach/2015/04/Close%20Calls%20with%20Nuclear%20Weapons.pdf
34. Krepon, M., 2009. *Better safe than sorry: The ironies of living with the bomb*. Stanford University Press.
35. Maloney, S., 2014. The missing essential part: Emergency provision of nuclear weapons for RCAF Air Defence Command, 1961–1964. *Canadian Military History* 23(1): 3.
36. Frankel, M., 2018. Learning from the missile crisis. In I.L. Horowitz (ed.) *Cuban Communism, 1959–2003*. Routledge, pp. 80–90.
37. Blight, J.G., Nye, J.S. and Welch, D.A., 1987. The Cuban missile crisis revisited. *Foreign Affairs* 66(1): 170–188.
38. Sagan, S.D., 1995. *The limits of safety: Organizations, accidents, and nuclear weapons*. Princeton University Press.
39. Union of Concerned Scientists, Close calls with nuclear weapons.
40. Blight, J.G. and Welch, D.A., 1995. Risking 'the destruction of nations': Lessons of the Cuban missile crisis for new and aspiring nuclear states. *Security Studies* 4(4): 811–850.
41. Future of Life Institute, Accidental nuclear war: A timeline of close calls: https://futureoflife.org/background/nuclear-close-calls-a-timeline/?cn-reloaded=1
42. White, M., 1995. *The Cuban Missile Crisis*. Springer.
43. George, A.L., 2004. *Awaiting Armageddon: How Americans faced the Cuban missile crisis*. UNC Press Books.
44. Sagan, *The limits of safety*.
45. Thayer, B.A., 1994. The risk of nuclear inadvertence: A review essay. *Security Studies* 3(3): 428–493.
46. Fairley, P., 2004. The unruly power grid. *IEEE Spectrum* 41(8): 22–27.
47. Megara, J., 2006. Dropping nuclear bombs on Spain: The Palomares accident of 1966 and the US Airborne Alert. MA thesis, Florida State University.
48. Montero, P.R. and Sánchez, A.M., 2001. Plutonium contamination from accidental release or simply fallout: study of soils at Palomares (Spain). *Journal of Environmental Radioactivity* 55(2): 157–165.

49. Stiles, D., 2005. A fusion bomb over Andalucia: US information policy and the 1966 Palomares incident. *Journal of Cold War Studies* 8(1): 49–67.
50. Center for Defense Information, 1981. U.S. nuclear weapons accidents: Danger in our midst. *The Defense Monitor* X(5): https://fas.org/nuke/norris/nuc_81010001a_n22.pdf
51. Phillips, A.F., 2001. Forgotten dangers of the Cold War: Nuclear accidents and nuclear winter. *Peace Research* 33(2): 129–144.
52. Pry, P.V. and McFerran, D., 1999. *War scare: Russia and America on the nuclear brink*. Greenwood Publishing Group.
53. Muñoz, C., 2010. Review under way: DOD inquiry into ICBM incident will focus on command and control issues. *Inside Missile Defense* 16(22): 5–6.
54. Schlosser, E., 2013. *Command and control: Nuclear weapons, the Damascus accident, and the illusion of safety*. Penguin.
55. BBC, 2018. Hawaii missile false alarm triggers shock, blame and apologies. 14 January: www.bbc.co.uk/news/world-us-canada-42680070
56. Acheson, R., 2018. The Doomsday machine: Confessions of a nuclear war planner. *Global Change, Peace & Security* 30(2): 278–282.
57. Deitchman, S., Dallas, C.E. and Burkle, F., 2018. Lessons from Hawaii: A blessing in disguise. *Health Security* 16(3): 213–215.
58. Freeman, A., 2018. Life, death and politics in Hawaii: 125 years of colonial rule. *The Conversation*, 17 January: https://theconversation.com/life-death-and-politics-in-hawaii-125-years-of-colonial-rule-90273
59. Wellerstein, A. 2018. The Hawaii alert was an accident, the dread it inspired wasn't. *Washington Post*, 16 January: www.washingtonpost.com/news/posteverything/wp/2018/01/16/the-hawaii-alert-was-an-accident-the-dread-it-inspired-wasnt/?utm_term=.610b770d4e54
60. Gregory, D., 2011. The everywhere war. *The Geographical Journal* 177(3): 238–250.
61. Leipnik, M.R., 2007. Use of geographic information systems in cyber warfare and cyber counterterrorism. In L. Janczewski and A. Colarik (eds), *Cyber warfare and cyber terrorism*. IGI Global, pp. 291–297.
62. Openshaw, S., Forbes, G.S., Miller, E.W. and Schmalz, R.F., 1992. The safety and siting of nuclear power plants when faced with terrorism and sabotage. In S.K. Majumdar (ed.), *Natural and technological disasters: Causes, effects and preventive measures*. Pennsylvania Academy of Science, pp. 455–468.
63. Cimbala, S.J., 2017. Nuclear deterrence and cyber warfare: Coexistence or competition? *Defense & Security Analysis* 33(3): 193–208.
64. NTI, 2015. The nuclear threat, 31 December: www.nti.org/learn/nuclear/
65. Nuclear Security Summit, 2016. Washington: www.nss2016.org
66. IAEA, Convention on the Physical Protection of Nuclear Material: www-ns.iaea.org/conventions/physical-protection.asp
67. The Global Initiative to Combat Nuclear Terrorism: www.gicnt.org
68. Parikh, N. et al., 2016. A comparison of multiple behavior models in a simulation of the aftermath of an improvised nuclear detonation. *Autonomous Agents and Multi-Agent Systems* 30(6): 1–27.

69. Walker, W., 1992. Nuclear weapons and the former Soviet republics. *International Affairs* 68(2): 255–277.
70. NTI, The nuclear threat: https://www.nti.org/
71. Lusher, A., 2017. Nuclear submarine sex and drugs scandal: Nine Trident crew expelled from Navy amid 'cocaine' and affairs allegations. 28 October: www.independent.co.uk/news/uk/home-news/navy-nuclear-submarine-sex-drugs-cocaine-hms-vigilant-trident-vanguard-women-in-armed-forces-a8024506.html
72. CBS, 2012. Navy submarine commander faked death to end affair with mistress. 18 September: www.cbsnews.com/news/navy-submarine-commander-faked-death-in-order-to-end-affair-with-mistress/
73. Ellison, A., 2017. Nuclear submarine sailor 'stole from prostitute'. 30 October: www.thetimes.co.uk/article/nuclear-submarine-sailor-stole-from-prostitute-nb8sh8nwc
74. BBC, 2011. Sailor who murdered officer on submarine HMS Astute jailed for life. 19 September: www.bbc.co.uk/news/uk-england-14971198
75. Sample, I., 2017. Treaty banning nuclear weapons approved at UN. *The Guardian*, 7 July: www.theguardian.com/world/2017/jul/07/treaty-banning-nuclear-weapons-approved-un
76. Beser, A., 2017. 122 countries have moved to ban nuclear weapons: What happens next? *National Geographic*, 7 July: https://blog.nationalgeographic.org/2017/07/07/122-countries-have-moved-to-ban-nuclear-weapons-what-happens-next/
77. ICAN, 2017. Poised to outlaw nuclear weapons. 6 July: www.icanw.org/campaign-news/poised-to-outlaw-nuclear-weapons/
78. Krebs, R., 2017. Why the Nobel Peace Prize brings little peace. *The Conversation*, 6 October: https://theconversation.com/why-the-nobel-peace-prize-brings-little-peace-84758
79. Wolfsthal, J., 2017. 1st nuclear ban draft is out. Arms Control Wonk, 22 May: www.armscontrolwonk.com/archive/1203255/the-1st-nuclear-ban-draft-is-out/
80. Wright, T. 2017. Nuclear ban treaty progresses, despite US-led objections. *The Interpreter*, 31 May: www.lowyinstitute.org/the-interpreter/nuclear-ban-treaty-progresses-despite-us-led-objections

8 Spaces of Peace

1. Galtung, J., 1959. Pacifism from a sociological point of view. *Journal of Conflict Resolution* 3(1): 67–84.
2. Morgan, P.M., 1983. *Deterrence: A conceptual analysis*. Sage Publications.
3. Brock, P. and Young, N., 1999. *Pacifism in the twentieth century*. Syracuse University Press.
4. DeLoughrey, E.M., 2013. The myth of isolates: Ecosystem ecologies in the nuclear Pacific. *Cultural Geographies* 20(2): 167–184. The nuclear cycle causes social justice issues from the moment uranium is extracted. This

is not my area of expertise, but I would recommend the brilliant work of Professor Gabrielle Hecht and Dr Stephanie Malin for those who want to learn more.

5. Thorpe, C., 2004. Against time: Scheduling, momentum, and moral order at wartime Los Alamos. *Journal of Historical Sociology* 17(1): 31–55.
6. Ibid.
7. Mian, Z., 2015. Out of the nuclear shadow: Scientists and the struggle against the Bomb. *Bulletin of the Atomic Scientists*, 71(1): 59–69.
8. Wang, J., 1999. *American science in an age of anxiety: Scientists, anticommunism, and the cold war*. University of North Carolina Press.
9. Yavenditti, M.J., 1974. The American people and the use of atomic bombs on Japan: The 1940s. *Historian* 36(2): 224–247.
10. Brown, A., 2012. *Keeper of the nuclear conscience: The life and work of Joseph Rotblat*. Oxford University Press.
11. Laucht, C., 2017. Transnational professional activism and the prevention of nuclear war in Britain. *Journal of Social History* 52(2): 439–467: https://doi.org/10.1093/jsh/shx032
12. Greco, P., 2017. Albert Einstein, pacifist. *Lettera Matematica* 5(1): 65–70.
13. Kraft, A., 2018. Dissenting scientists in early Cold War Britain: The 'fallout' controversy and the origins of Pugwash, 1954–1957. *Journal of Cold War Studies* 20(1): 58–100.
14. Laucht, Transnational professional activism.
15. Ibid.
16. https://thebulletin.org is the current place to see the Doomsday Clock and learn more about BAS.
17. Mcdonald, J.C., 2017. 'Widening the web'. Greenham Common, the CND and the Women's Movement: The rise and fall of women's antinuclear activism, 1958–1988. Master's thesis, University of Oslo: https://www.duo.uio.no/handle/10852/59932
18. St John, G., 2008. Protestival: Global days of action and carnivalized politics in the present. *Social Movement Studies* 7(2): 167–190.
19. Eschle, C., 2016. Faslane Peace Camp and the political economy of the everyday. *Globalizations* 0(0): 1–3: doi:10.1080/14747731.2016.1156321
20. Boyer, P., 1994. *By the bomb's early light*, University of North Carolina Press, p. 90. For more on the impact on Bikini, see Boyer, *By the bomb's early light*; Firth, S., 1987. *Nuclear playground*. University of Hawai'i Press. Niedenthal, J., 2001. *For the good of mankind: A history of the people of Bikini and their islands*. Bravo Publishers. Stone, R., 1988. *Radio Bikini*. Robert Stone Productions. Dibblin, J., 1988. *Day of two suns: Nuclear testing and the Pacific Islanders*. Virago. Teaiwa, T., 1994. Bikinis and Other s/pacific n/oceans. *The Contemporary Pacific* 6. Weisgall, J., 1994. *Operation crossroads: The atomic tests at Bikini Atoll*. Naval Institute Press.
21. Dundas, M., 2015. Founding a nonviolent community at Bangor, Washington: The Peace Blockade, Part 1. 31 July. Satyagraha Foundation for Nonviolence Studies: www.satyagrahafoundation.org/founding-a-nonviolent-community-at-bangor-washington-the-peace-blockade-part-1/

22. Intondi, V.J., 2015. *African Americans against the bomb: Nuclear weapons, colonialism, and the Black freedom movement*. Stanford University Press.
23. Nau, H.R., 2018. *At home abroad: Identity and power in American foreign policy*. Cornell University Press.
24. Jorgensen, T., 2016. Bikini islanders still deal with fallout of US nuclear tests, 70 years later. *The Conversation*, 29 June: https://theconversation.com/bikini-islanders-still-deal-with-fallout-of-us-nuclear-tests-70-years-later-58567
25. Cho, Y.F., 2018. Remembering Lucky Dragon, re-membering Bikini: Worlding the Anthropocene through transpacific nuclear modernity. *Cultural Studies* 33(1): 1–25.
26. Check out Kathleen Sullivan and her work with *Hibakusha* in New York: www.hibakushastories.org
27. Lamar Jr., J.V., Aikman, D. and Amfitheatrof, E., 1985. Another return from the cold. *Time*, 7 October.
28. Dawson, J.I., 1996. *Eco-nationalism: Anti-nuclear activism and national identity in Russia, Lithuania, and Ukraine*. Duke University Press.
29. Gerasimov, I.P., 1985. Geography of peace and war: A Soviet view. In D. Pepper and A. Jenkins (eds) *The geography of peace and war*, Blackwell, p. 201.
30. Gusterson, H., 1999. Los Alamos: Summer under siege. *Bulletin of the Atomic Scientists* 55(6): 36-41.
31. Jones, C., 2006. Ed Grothus and the Doomsday Stones How a former nuclear-weapons machinist and self-appointed cardinal of the First Church of High Technology in Los Alamos, New Mexico, intends to save the world. *Esquire-New York* 145(4): 172.
32. For more about current affairs of the WPC see: www.wpc-in.org
33. For or more about CND see: www.cnduk.org/
34. London School of Economics and Warwick University Modern Records store CND archival material. https://archiveshub.jisc.ac.uk/search/archives/4526124c-662d-336e-8eec-28b2c173aeac
35. Hudson, K., 2005. *CND – now more than ever: The story of a peace movement*. Summersdale Publishers.
36. Hill, C.R., 2018. *Peace and power in Cold War Britain: Media, movements and democracy, c. 1945–68*. Bloomsbury.
37. Matheson, M.J., 1997. The opinions of the International Court of Justice on the threat or use of nuclear weapons. *American Journal of International Law* 91(3): 417–435.
38. Hawkins, D. Ruff, T., 2017. How Melbourne activists launched a campaign for nuclear disarmament and won a Nobel Prize. *The Conversation*, 19 October: https://theconversation.com/how-melbourne-activists-launched-a-campaign-for-nuclear-disarmament-and-won-a-nobel-prize-85386
39. Bolton, M. and Minor, E., 2016. The discursive turn arrives in Turtle Bay: The international campaign to abolish nuclear weapons' operationalization of critical IR theories. *Global Policy* 7(3): 385–395.

40. Wright, T., 2018. ICAN: The beginning of the end of nuclear weapons symposium, 22 July, Hiroshima, Japan.
41. Ibid.
42. Wright, T., 2017. Nuclear ban treaty progresses, despite US-led objections. *The Interpreter* (The Lowy Institute), 31 May: www.lowyinstitute.org/the-interpreter/nuclear-ban-treaty-progresses-despite-us-led-objections
43. Ibid.
44. Laforge, J., 2017. Activists challenge US nukes in Germany: Occupy bunker deep inside nuclear weapons base. *Counterpunch*, 21: www.counterpunch.org/2017/07/21/activists-challenge-us-nukes-in-germany-occupy-bunker-deep-inside-nuclear-weapons-base/
45. DW, 2017. The last nukes in Germany: www.dw.com/en/the-last-nukes-in-germany/a-18630943
46. International Action Camp Review of 2017, 10–18 July 2018: https://buechel-atombombenfrei.jimdo.com/international/international-action-camp/
47. Hodder, J., 2017. Waging peace: Militarising pacifism in Central Africa and the problem of geography, 1962. *Transactions of the Institute of British Geographers* 42(1): 29–43.

9 Future War Zones

1. Kristian, H.M. and Norris, R.S., 2018. Status of world nuclear forces. Federation of American Scientists: https://fas.org/issues/nuclear-weapons/status-world-nuclear-forces/
2. Lowenthal, M., 2018. Nuclear disarmament and non-proliferation today and in the near future. In *International cooperation for enhancing nuclear safety, security, safeguards and non-proliferation: 60 years of IAEA and EURATOM. Proceedings of the XXth Edoardo Amaldi Conference, Accademia Nazionale dei Lincei, Rome, Italy, 9–10 October 2017*. Springer, pp. 195–202.
3. Gibbons, R.D., 2018. The humanitarian turn in nuclear disarmament and the Treaty on the Prohibition of Nuclear Weapons. *The Nonproliferation Review* 25(1): 11–36.
4. Podvig, P., 2016. Blurring the line between nuclear and nonnuclear weapons: Increasing the risk of accidental nuclear war? *Bulletin of the Atomic Scientists* 72(3): 145–149.
5. Kahn, H. and Jones, E., 2017. *On thermonuclear war*. Routledge.
6. Thakur, R., 2018. Japan and the Nuclear Weapons Prohibition Treaty: The wrong side of history, geography, legality, morality, and humanity. *Journal for Peace and Nuclear Disarmament*, 1(1): 11–31.
7. Treaty for the Prohibition of Nuclear Weapons in Latin America (Tlatelolco Treaty). IAEA: www.iaea.org/publications/documents/treaties/treaty-prohibition-nuclear-weapons-latin-america-tlatelolco-treaty
8. Hamel-Green, M., 1998. The South Pacific – The Treaty of Rarotonga. In R. Thakur (ed.), *Nuclear weapons-free zones*. Palgrave Macmillan, pp. 59–80.

9. Robie, D., 1986. *Eyes of fire: The last voyage of the Rainbow Warrior*. Lindon Publishing.
10. Thakur, R., 1986. A dispute of many colours: France, New Zealand and the Rainbow Warrior affair. *The World Today* 42(12): 209–214.
11. BBC, 2005. French expats recall NZ bombing. 8 July: http://news.bbc.co.uk/1/hi/world/asia-pacific/4637897.stm
12. Robie, D., 2006. Revisiting French terrorism in the Pacific: Rainbow Warrior revelations. *Pacific Ecologist* 12: 26–29.
13. Roscini, M., 2008. Something old, something new: The 2006 Semipalatinsk Treaty on a nuclear weapon-free zone in Central Asia. *Chinese Journal of International Law* 7(3): 593–624.
14. Enkhsaikhan, J., 2017. Promoting Mongolia's nuclear-weapon-free status: Lessons learned and relevance for Northeast Asia, APLN/CNND Policy Brief. No. 43. Asia–Pacific Leadership Network: www.a-pln.org/briefings/briefings/
15. African Nuclear Weapon-Free Zone Treaty (Pelindaba Treaty). IAEA: www.iaea.org/publications/documents/treaties/african-nuclear-weapon-free-zone-treaty-pelindaba-treaty
16. Stockholm International Peace Research Institute, 2005. *SIPRI yearbook 2005: Armaments, disarmament, and international security*. SIPRI.
17. Borger, J. and Sample, I., 2018. All you wanted to know about nuclear war but were too afraid to ask. *The Guardian*, 16 July: www.theguardian.com/world/2018/jul/16/nuclear-war-north-korea-russia-what-will-happen-how-likely-explained
18. Purkitt, H.E. and Burgess, S.F., 2005. *South Africa's weapons of mass destruction*. Indiana University Press.
19. Walker, W., 1992. Nuclear weapons and the former Soviet republics. *International Affairs* 68(2): 255–277.
20. Borger and Sample, All you wanted to know about nuclear war but were too afraid to ask.
21. Liechtenstein, S., 2017. OSCE structured dialogue: Countering the risk of military escalation in Europe. *Security and Human Rights Monitor*, 19 November:www.shrmonitor.org/osce-structured-dialogue-countering-risk-military-escalation-europe/
22. Power, P.F., 1986. The South Pacific Nuclear-Weapon-Free Zone. *Pacific Affairs* 59(3): 475.
23. Freedman, L., 2013. Disarmament and other nuclear norms. *The Washington Quarterly* 36(2): 93–108.
24. Rublee, M.R., 2009. *Nonproliferation norms: Why states choose nuclear restraint*. University of Georgia Press.
25. Reinventing Civil Defense is a project at the College of Arts and Letters at the Stevens Institute of Technology, funded by the Carnegie Corporation of New York, running from 2017 to 2019: https://reinventingcivildefense.org/project/

26. Preston, J., 2008. Protect and Survive: 'Whiteness' and the middle-class family in civil defence pedagogies. *Journal of Education Policy* 23(5): 469–482.
27. Andy Crooks research profile and project page: www.researchgate.net/project/Individual-and-social-response-to-a-nuclear-WMD-event
28. Desai, S., Bell, W., Harris, C. and Dallas, C., 2018. Human consequences of multiple nuclear detonations in New Delhi, India: www.preprints.org/manuscript/201806.0438/download/final
29. Lazaroff, C. 2018. Dawn of a new Armageddon. *Bulletin of the Atomic Scientists* (Pre-pub).
30. Long, A., 2018. *Russian nuclear forces and prospects for arms control*. RAND Corporation.
31. U.S. Department of Defense, 2018. *Nuclear posture review*. Washington, DC, p. 8.
32. Long, *Russian nuclear forces and prospects for arms control*.
33. U.S. Department of Defense, *Nuclear posture review*, p. 8.
34. BBC. 2018. Trump sides with Russia against FBI at Helsinki Summit. 16 July: www.bbc.co.uk/news/world-europe-44852812
35. Ibid.
36. BBC. 2018. Trump-Russia affair: Key questions answered. 13 July: www.bbc.co.uk/news/world-us-canada-42493918
37. Davidson, A., 2018. The theory of Trump compromat. *The New Yorker*, 19 July: www.newyorker.com/news-desk/swamp-chronicles/a-theory-of-trump-kompromat
38. Thakur, R., 2016. Nuclear stability in Asia, strengthening order in times of crisis. 10th Berlin Conference on Asian Security (BCAS), 19–21 June. www.swp-berlin.org/fileadmin/contents/products/projekt_papiere/BCAS2016_Ramesh_Thakur_web.pdf
39. O'Neil, A., 2013. *Asia, the US and extended nuclear deterrence: Atomic umbrellas in the twenty-first century*. Routledge.
40. Ibid.
41. Rublee, *Nonproliferation norms*.
42. *New York Times*, 2018. Trump abandons Iran nuclear deal he long scorned. 8 May: www.nytimes.com/2018/05/08/world/middleeast/trump-iran-nuclear-deal.html
43. *The Atlantic*, 2018. This is bigger than a meeting with Kim Jong-Un. June: www.theatlantic.com/international/archive/2018/06/trump-kim-summit/562346/
44. BBC, 2018. Three reasons behind Trump ditching the Iran deal. 8 May: www.bbc.co.uk/news/world-us-canada-43902372
45. *The Independent*, 2018. The appointments of John Bolton and Mike Pompeo in the US bring us closer to war in the Middle East. 30 March: www.independent.co.uk/voices/john-bolton-donald-trump-mike-pompeo-middle-east-iran-a8281436.html
46. Tayebipour, M., 2018. Netanyahu's attempt to discredit the Iran nuclear deal doesn't hold water. *The Conversation*. 2 May: https://theconversation.

com/netanyahus-attempt-to-discredit-the-iran-nuclear-deal-doesnt-hold-water-95829
47. *UK Cabinet Submission from Joint Intelligence Bureau.* Cabinet Office, Government of the United Kingdom. 27 March 1961. JIC/519/61: http://webarchive.nationalarchives.gov.uk/20070205123750/; www.cabinetoffice.gov.uk/foi/pdf/israeli_nuclear2.pdf
48. Kroenig, M., 2009. Importing the bomb: Sensitive nuclear assistance and nuclear proliferation. *Journal of Conflict Resolution* 53(2): 161–180.
49. Turner, S., 2018. *Caging the genies: A workable solution for nuclear, chemical, and biological weapons.* Routledge.
50. White House, 2018. Remarks by President Trump and Prime Minister Abe of Japan before bilateral meeting. 7 June: www.whitehouse.gov/briefings-statements/remarks-president-trump-prime-minister-abe-japan-bilateral-meeting-3/
51. *The Telegraph,* 2018. Dennis Rodman with Kim Jong-un in North Korea. 14 January: www.telegraph.co.uk/news/picturegalleries/worldnews/9902493/Dennis-Rodman-with-Kim-Jong-un-in-North-Korea.html?frame=2496698
52. Davis, T.C., 2007. *Stages of emergency: Cold War nuclear civil defense.* Duke University Press.
53. Standing Senate Committee on National Security and Defence, Canada, 2014. *Canada and ballistic missile defense: Responding to the evolving threat*: https://sencanada.ca/content/sen/committee/412/secd/rep/rep10jun14-e.pdf
54. *National Post,* 2017. Why Canada would be directly in the way of a North Korean nuclear war. 3 August: https://nationalpost.com/news/canada/why-canada-would-be-directly-in-the-way-of-a-north-korean-nuclear-war
55. Pollack, J., 2017. Nuclear deterrence and the revenge of geography. Arms Control Wonk: www.armscontrolwonk.com/archive/1204122/nuclear-deterrence-the-revenge-of-geography/
56. NTI, 2011. SALT II: www.nti.org/learn/treaties-and-regimes/strategic-arms-limitation-talks-salt-ii/
57. Ganguly, S. and Hagerty, D.T., 2012. *Fearful symmetry: India–Pakistan crises in the shadow of nuclear weapons.* University of Washington Press.
58. *The Wire,* 2017. Nuclear Ban Treaty doesn't contribute to customary international law: India. 18 July: https://thewire.in/159057/nuclear-ban-treaty-customary-law/?mkt_tok=
59. Jo, D.J. and Gartzke, E., 2007. Determinants of nuclear weapons proliferation. *Journal of Conflict Resolution* 51(1): 167–194.
60. Hundley, T., 2018. India and Pakistan are quietly making nuclear war more likely. *Vox,* 4 April: www.vox.com/2018/4/2/17096566/pakistan-india-nuclear-war-submarine-enemies
61. STRATFOR, 2018. Pakistan calls India's nuclear bluff in a subcontinent standoff. 4 April: https://worldview.stratfor.com/article/pakistan-india-nuclear-standoff-subcontinent
62. Hundley, India and Pakistan are quietly making nuclear war more likely.

63. Paul, T.V. and Shankar, M., 2007. Why the US–India nuclear accord is a good deal. *Survival* 49(4): 111–122.
64. Jo and Gartzke, Determinants of nuclear weapons proliferation.
65. Robock, A. and Toon, O.B., 2012. Self-assured destruction: The climate impacts of nuclear war. *Bulletin of the Atomic Scientists* 68(5): 66–74.
66. Toon, O.B., Robock, A., Mills, M. and Xia, L., 2017. Asia treads the nuclear path, unaware that self-assured destruction would result from nuclear war. *Journal of Asian Studies* 76(2): 437–456.
67. Robock, A. and Toon, O.B., 2016. We still have enough nuclear weapons to cause a nuclear winter. *The New York Times*, 11 February: www.nytimes.com/2016/02/11/opinion/lets-end-the-peril-of-a-nuclear-winter.html?smid=nytcore-iphone-share&smprod=nytcore-iphone
68. MacKenzie, D. and Spinardi, G., 1995. Tacit knowledge, weapons design, and the uninvention of nuclear weapons. *American Journal of Sociology* 101(1): 44–99.
69. UN Secretary-General, 2018. *Securing our common future: An agenda for disarmament*. UNODA. Sales No. E.18.IX.6: www.un.org/disarmament/publications/more/securing-our-common-future/
70. UK Government, 2018. Consultation on draft nuclear safeguard regulations: https://beisgovuk.citizenspace.com/civil-nuclear-resilience/nuclear-safeguards-regulations/
71. Mason, R. and Asthana, A., 2018. Commons votes for Trident renewal by majority of 355. *The Guardian*, 18 July: www.theguardian.com/uk-news/2016/jul/18/mps-vote-in-favour-of-trident-renewal-nuclear-deterrent
72. Richie, R., 2016. Why British politicians find it so hard to vote against nuclear weapons. *The Conversation*, 19 July: https://theconversation.com/why-british-politicians-find-it-so-hard-to-vote-against-nuclear-weapons-62655

Index

A-bomb dome 23, 27
Abon, Lemyo 59
activism
  by academics 7
  extensive history of 6
  range of activities 106
  see also pacifists and pacifism
acute radiation syndrome (ARS) 15–16, 18, 87, 113
Afghanistan 4
Africa, nuclear-free zone in 127 see also Global South
African-Americans 111–13
agnotology 15
Algerian Sahara 55, 61
Anti-Ballistic Missile Treaty 130
Argentina 127
Arkhipov, Vasili 98
arms control treaties see non-proliferation treaties
Askins, Kye 7
Atomic Scientists Association (ASA) 108
atomic veterans
  families of 42–3, 48–50
  health and safety environment for 30–1, 35, 37, 39–40
  health impacts for: difficulty of measuring 31, 43–7; perceptions of 35, 41–3, 47, 48–50
  memorialisation by 50–2, 71
  number of 34tab
  support for 47–50
Attlee, Clement 118
Australia
  atomic veterans of 35, 37, 45, 51
  nuclear testing in 35–6, 54, 55–8
  nuclear weapons policy of 123
authoritarianism, rise of 93

Baaro, Makurita 60
Baglay, Gopal 136
Ban Treaty (UN Treaty on the Prohibition of Nuclear Weapons, 2017)
  Australia's rejection of 123
  ICAN and 6
  India's rejection of 136
  Japan's rejection of 128
  non-nuclear states, agency for 92
  nuclear de-normalisation and 96–7, 122
  potential challenges of 103, 129
  status of 95tab, 104–5, 122, 126
  see also Comprehensive Test Ban Treaty (1996); Partial Test Ban Treaty (1963)
Begin Doctrine 133
Belarus 127
Ben-Gurion, David 133
Bikini Atoll 32, 54, 55, 59–60, 113
biomedical disease labels 46
biopolitical endpoints 46
birth defects 49, 63, 66
Black Hole of Los Alamos 116–18
'black mist' (*puyu*) 56–7
BNTVA (British Nuclear Test Veterans Association) 44–5, 48, 50–2
Born, Max 108

Boyer, Paul 111
Brazil 127
Brexit 137
Bross, I.D. and N.S. 43–4
Brown, Kate 6–7
Brumby (Operation) 36, 57
Büchel Airbase invasion 123–4
Bulletin of the Atomic Scientists (BAS) 109

INDEX

Bunge, 'Wild Bill' 79, 81–2
bunkers 68, 72–3, 75–6, 77, 87–8, 89
Burchett, Wilfred 15

Canada 134
cancer 44, 45, 61, 63, 66
cardiovascular diseases 63
cartographies 82–4
censorship *see* propaganda and information suppression
Central Asian nuclear-free zone 127
Certeau, Michel de 1
Channel 4 66
China 1–2, 55, 64–6, 132
Christmas Island (Kiritimati Island) 32, 33, 36–41, 55, 60–1, 71
Church of High Technology 116–17, 118
Churchill War Rooms 68, 72
civil defence 74–6, 82, 129, 134 *see also* 'preppers'
clean-up operations 36, 48, 54, 57, 60
CND (Campaign for Nuclear Disarmament) 110, 119–21
colonialism
    Black activism and 112–13
    colonies as test sites 31, 32, 36–41, 53–62
'comfort women' 22, 27
compensation
    lack of 37, 58, 66
    provision of 37, 48, 62
Comprehensive Test Ban Treaty (CTBT, 1996) 95tab, 107
computer games 79–80, 84–90
concealment/disguising of sites 1–3, 71–2
Corbyn, Jeremy 5
CPPNM (Convention on the Physical Protection of Nuclear Material) 102
critical scenario studies 129
Crooks, Andy 130
Cuban Missile Crisis 97–9, 107
Cutter, S.L. 81
cyber attacks 101–2

Daghlian, Harry 18
DDT 39–40
Death on the Silk Road (documentary) 66
Defence Threat Reduction Agency study 129–30
dehumanisation of civilians 14, 53, 55, 58, 104
DeLillo, Don 68
demon core 16–18
deterrence norm 91
disarmament 6, 96, 127–8
displacement of locals
    Australia 54, 55, 56–8
    China 55
    French colonies 55, 61
    South Pacific 54, 55, 58–62
    US 55
    USSR 32–3, 55, 62
Dominic (Operation) 36
Donovan, Ryan Samuel 104
Doomsday Clock 109
Doomsday Stones 117–18

East Turkestan (Xianjiang) 65–6
ECAS (Emergency Committee of Atomic Scientists) 108
ecofeminism 5
Ehlers, Freda 120
Einstein, Albert 108
election interference 131–2
electromagnetic pulse (EMP) 101–2
Emu Field 55–6
Eniwetok Atoll 32, 55
Entewak Atoll 54, 60
extremist groups 28–9

Fallout (computer game) 85, 86–90
Fangataufa Atoll 55, 61
Faslane Peace Camp 111
'feral ghouls' 88
Fihn, Beatrice 6
Fiji 62
'first-use' policies 132
fishermen, radiation exposure by 32
Fission Line 48

framing *see* propaganda and
    information suppression
France 33, 51, 126–7
Fukushima disaster 113

Galtung, Johan 106
game theory 80
gaming 79–80, 84–90
Gandhi, Mahatma 106
Garcia, B. 41
genetic impacts 35, 49–50, 54, 63
geographers, as pacifists 80–2,
    114–15, 137–8
geographical engineering 80
geotechnologies 82–4, 101–2, 130
Gerasimov, I.P. 114–15
Germany 11
'ghouls' 88
Global Initiative to Combat Nuclear
    Terrorism 102
Global North 94–7, 128, 136 *see also*
    nuclear weapons possessor states
Global South, nuclear-free zones in
    105, 107, 121–2, 126–7
Golders Green Guildswomen 120
Gorbachev, Mikhail 6, 93
government culpability 46
Grapple (Operations) 36–41, 60–1
Greenham Common peace camp
    109–10
Greenland, false alarm in 97
Greenpeace 126–7
Grothus, Ed 9, 115–18
Guterres, António 136

Hanford nuclear laboratories 10
'Hard Rock' exercise 76
Hawaiian false missile alert 99–101
health and safety environment
    for atomic veterans 30–1, 35, 37,
        39–40
    for locals 56–7, 60–5
health effects and deaths
    atomic veterans 35–50; difficulty of
        measuring 31, 43–7; perceptions
        of 35, 41–3, 47, 48–50

global deaths from testing 53
Hiroshima and Nagasaki 20–2,
    26–7
local populations 54, 56–7, 61,
    62–3, 66, 113
scientific workers 16–18
healthy soldier effect 43
Helsinki Summit 131
Heritage Foundation 114
Hersey, John 15
Hibakusha 9, 22, 26–7, 29, 113–14
Hiroshima
    attack on 9, 10–16
    cartographies of 83
    effects of bombing 20–2, 26–7
    memorialisation in 22–3, 25–9
    Obama's visit to 96
    reconstruction in 21–3, 25, 27
Hiroshima Peace Memorial Park 23,
    27–9
Hopkins, Sarah 110
horizontal proliferation 91
Hornig, Lilli 19
hydrogen bombs 33, 75, 97

ICAN (International Campaign to
    Abolish Nuclear Weapons) 6,
    104–5, 109, 121, 122–3
'ICBMs', as obfuscating term 3
India 8, 130, 132, 135–6
indigenous people *see* local
    communities
inequality
    after bombings of Japan 22
    bunkers for elites only 72, 74, 75–6,
        77
    in computer games 87–8, 89
    Los Alamos and surrounding area
        24–5
    nuclear colonialism 53–67
    US vs Japan 25
infertility 49
information suppression
    on atomic bombing of Japan 15–16
    on atomic testing 32, 41, 42
insider crimes 103–4

Institute of Radiation Safety and
    Ecology (IRSE) 63
Intermediate-range Nuclear Forces
    Treaty 6
International Court of Justice 60, 122
Intondi, Vincent 112–13
ionising radiation 15–16, 17–18, 22
    atomic veterans' exposure to 41–7;
        families of vets 48–50
    global circulation of 62
    locals' exposure to 31, 32–3, 56–7,
        61–7, 113
IPPNW (International Physicians for
    the Prevention of Nuclear War)
    53, 109
Iran 8, 132–3
Iraq 127–8
Israel 76, 133
Iversen, Kristen 7

Japan
    equality in 22, 25
    memorialisation in 22–3, 25–9
    pacifism in 113–14
    rejection of Ban Treaty by 128
    US alliance and 132
    see also Hiroshima; Nagasaki
Japanese fishermen, radiation
    exposure by 32
Jaspal, Zafar 136
Jenkins, A. 6
Johnston Atoll 36, 55
Johnstone, Phil 7

Kahn, Herman 85
Kalama Atoll 32, 55
Kapustin Yar 55
Karle, Isabella 19
Kazakhstan 32–3, 54, 55, 62–4, 127
Keane, John 57
Kennedy, John F. 96, 98, 125
Khan, Herman 93
Khrushchev, Nikita 97–8
Kim Jong-un 5, 8, 93, 133–4
King, Martin Luther, Jr 112

Kiritimati Island (Christmas Island)
    32, 33, 36–41, 55, 60–1, 71
Kokura, Japan 11–12
Koreans, in Japan 22, 27
Kwajalein Atoll 59
Kyoto, Japan 13
Kyrgyzstan 127

landscapes 69–74
Lange, David 121–2
language (euphemisms) 3–4 *see also*
    propaganda and information
    suppression
LANL *see* Los Alamos
Lashmar, Paul 13
leaflet drop, on Hiroshima 14
leukaemia 43, 44, 45, 61, 63, 66
Libya 127–8
local communities 53–67
    health effects for 54, 56–7, 61,
        62–3, 66, 113
    health impacts for 54, 56–7, 61,
        62–3, 66, 113
    as human guinea pigs 31, 33, 62–3,
        64
    ionising radiation, exposure to 31,
        32–3, 56–7, 61–7, 113
    lack of support for 57, 61–2, 64, 66
    see also displacement
Lop Nor 55, 64–6
Loreak, Christopher 62
Los Alamos (US)
    nuclear complex at 2, 9, 10–12,
        17–20, 24–5
    pacifism at 115–18
    tourist park at 71–2, 73tab
low-yield nuclear weapons 131
Lucky Dragon fishing ship 113
ludology 84

MAD (mutually assured destruction)
    80, 85, 93, 96
Mafart, Alain 127
Malden Island 33, 60
Malin, Stephanie 7
Manchester 76

Manhattan Project 10, 17, 19–20 *See also* Los Alamos
Manhattan Project National Historical Park 71–2, 73fig
Manhattan Project scientists, as pacifists 107–8
maps, concealment by 1–3
Maralinga 33, 35–6, 54, 55–8
Maralinga Tours 58
Marshall Islands 32, 55, 58–60, 62
Marshall, Leona Woods 19
Mason, Kelvin 7
materialities of nuclear warfare 69
May, Theresa 5, 137
Mayer, Maria Goeppert 19
McClelland Royal Commission 57
medical records 41
Medvedev, Dmitri 96
Meitner, Lisa 20
Mercury secret city (US) 2
militarisation, rise of 93
military, the, crimes within 103–4
Milligan, Spike 119
Million Persons study 45, 48
miscarriages 35, 49
Missile Command (computer game) 85
mistakes and near misses 97–101
MOAB (Massive Ordnance Air Blast) 4, 126
Molesworth People's Peace Camp 111
Mongolia 127
Montebello Island 33, 55
Mororua Atoll 55, 61
Muirhead, C.R. 44
multiple myeloma 44
Mulvihill, Michael 69
Mumford, Lewis 108
Murphy, B.C. 42
museums and heritage sites 68–9, 70, 71–2, 73fig
Muto Ichiyo 113

NAAV (National Association of Atomic Veterans) 48
Nagasaki
 attack on 9, 10–16
 cartographies of 83
 effects of bombing 20–2
 memorialisation in 26
narratives *see* propaganda and information suppression
National Command Level (NCL) 85
NATO 130
Nazis 11
near misses and false alarms 97–101
necropolitics 67
neoliberalisation of civil defence 75, 76–8
 in computer games 87–8, 89
Netanyahu, Benjamin 133
New Delhi 130
New Zealand 121–2
NewSTART 96, 130–1, 135
Nkrumah, Kwame 121
'no-first-use' policies 132
Nobel Peace Prize 96, 105, 108, 109
non-proliferation treaties 91, 94–7, 95tab, 104–5 *see also specific treaty*
Non-Proliferation Treaty (NPT, 1970) 3–4, 91, 95tab, 105, 131, 133
non-use norm 91
North Korea 8, 132, 133–5, 136 *See also* Kim Jong-un
Novaya Zemlya 32–3, 55, 62
NPT *see* Non-Proliferation Treaty
nuclear bunkers 68, 72–3, 75–6, 77, 87–8, 89
nuclear colonialism/imperialism 53–67, 121
nuclear energy 113–14
Nuclear Families study 37–43
nuclear-free zones 75–6, 105, 107, 121–2, 126–8
nuclear norms 91
nuclear proliferation 8, 30, 125
'Nuclear Refrain' project 7
nuclear refugees 53–67
nuclear taboo 91–4, 122
nuclear terrorism 102–4
nuclear test veterans *see* atomic veterans

INDEX · 171

nuclear testing 30–7, 55–67 *see also* atomic veterans; local populations
nuclear tourism 54, 57–8, 68–74
Nuclear war atlas (Bunge) 79, 81–2
nuclear weapon possessor states 3–4, 8, 91–2, 94–7, 105, 126
nuclear winter 136
nuclearism 92
NukeMap 82

Obama, Barack 96
Oppenheimer, J. Robert 11, 16, 107–8
Organisation for Security and Co-operation in Europe 128
Orwell, George 24–5
Outer Space Treaty 84

Pacific Proving Grounds 36–41
pacifists and pacifism 80–2
　criminal behaviour by 123–4
　definition and range of 107
　effectiveness of 107
　geographers 80–2, 114–15, 137–8
　major NGOs 118–21 (*see also* ICAN)
　scientists 107–9
Pakistan 8, 132, 135–6
Pape, Robert 13
Parkinson, Alan 36
Partial Test Ban Treaty (1963) 95tab, 107
patriarchy 5
Pauling, Linus 108
peace camps 109–11
Pelindaba Treaty 127
Penney, William 30
Pepper, D. 6
Petrov, Stanislav 99
pilots 13
play *see* computer games
plutonium pit manufacturing plants 70
poetry 26
political influences 46

possessor states 3–4, 8, 91–2, 94–7, 105, 126
post-nuclear landscapes 69–74
Preece, Richard 57
'preppers' 76–8
Price, Tony 116
Prieur, Dominique 127
privatisation of civil defence 75–8
　in computer games 87–8, 89
Project Ploughshare 80
Project Y *see* Los Alamos
proliferation 93–4
propaganda and information suppression
　on atomic bombing of Japan 13–16, 14, 72
　on atomic testing 31, 32, 36, 41, 42, 53, 55, 58
　Cold War sites as opportunity for 71
　dehumanisation of civilians 14, 53, 55, 58, 104
　games as legitimising nuclear war 85
　on nuclear energy 113
　on nuclear-free zones 72
　on nuclear weapon possession 92, 94
　pacifism, erasure of 106
Pugwash Conferences 108–9
Putin, Vladimir 93, 131

Quakers 111, 119

racism
　dehumanisation of locals 14, 53, 55, 58, 104
　racist stereotypes 57, 60–1
　'uninhabited/wasteland' claims 31, 53, 55, 58
Radiation Dose Reconstruction Team 48
radical geography of nuclear warfare 5–8
Rainbow Warrior (ship) 127
RAND Corporation 85, 93

Rarotonga Island 60
Rarotonga, Treaty of 121, 126–7
Reagan, Ronald 6, 93
Reinventing Civil Defense project 129
relational materiality 69
reproductive health issues 35, 49, 63
Rocky Flats National Wildlife Refuge 70
Rodman, Dennis 134
Roff, S.R. 44–5
Rongelap Atoll 59–60
Roosevelt, Franklin D. 10, 79
Rotblat, Joseph 108
Roy, Arundhati 53
Runit Dome 54, 60
Russell, Bertrand 108, 119, 120
Russell-Einstein pacifist manifesto 108
Russia
   arms control efforts 96, 131
   election interference by 131–2
   pacifism and 114–15
   secret cities in 2
   see also Soviet Union
Ryan, Paul 131

Sagan, Carl 91
Sankichi Tōge 22–3
Saudi Arabia 132
Second World War 10–16
secrecy, culture of 1, 30, 42, 48, 63, 71–2
secret cities 2–3 *see also* Los Alamos
self-assured destruction (SAD) 136
Semipalatinsk, and Treaty of 32–3, 54, 62–4, 95tab, 127
simulations 85–6
Singapore 132
Slotin, Louis 18
Smart, Mima 56, 57
Smyth, Henry DeWolf 11–12
South Africa 127
South Korea 127, 132
Soviet Union
   arms control efforts 6
   civil defence in 74, 75
   secret cities in 2
   testing by 32–3, 55, 62–4

Spain, accidental bomb drop on 99
Speier, Hans 81
S.T.A.L.K.E.R. 86
Steele, Sheila and Harold 119
stigmatisation
   of nuclear weapons 93, 122
   of victims 16, 22, 54, 63, 88
Stimson, Henry L. 13
Suez Crisis 97
Sunao Tsuboi 22
surveillance 83
survivalists 76–8
Sweden 127
Switzerland 76
Symonds, J.K. 56–7
Szilárd Petition 107–8

taboo (nuclear) 91–4, 122, 131
Taiwan 127
Tajikistan 127
Takada Jun 66
Tanka poems 26
Tannenwald, Nina 93
terrorism 102–4
testing 30–7, 55–67 *see also* atomic veterans; local populations
Thatcher, Margaret 6, 93, 109
theft of nuclear materials 102, 103
Tlatelolco Treaty 95tab, 126
Totem I 56
tourism 54, 57–8, 68–74
treaties *see* non-proliferation treaties; *specific treaty*
Treaty on the Prohibition of Nuclear Weapons 4, 95tab
Trident 33
Trinity bomb 11
Truman, Harry S. 11
Trump, Donald
   deployment of MOAB by 4
   Kim Jong-un and 5, 8, 134
   Middle East and 132–3
   nuclear treaties, rejection of 6, 93, 96, 131
   Russia, dynamic with 131–2
   temperament issues of 125–6, 131

Trundle, C. 46
Tsar Bomba 33, 62
Tsutomu Yamaguchi 26
Turkmenistan 127

Ukraine 127
UN (United Nations) 58, 59, 60
UN Outer Space Treaty 84
UN Treaty on the Prohibition of Nuclear Weapons (2017) *see* Ban Treaty
under-urbanisation 3
United Kingdom
  atomic veterans of 34tab, 35–45, 47–8, 50–2
  civil defence in 74–6, 77, 82
  compensation by 37, 58
  crimes by military members 103–4
  pacifism in 108–11, 118–21
  testing by 33–41, 54, 55–8, 60–1
  uncertainty in 137
United States
  arms reduction efforts 6, 93, 96
  atomic bombing of Japan 9–16
  atomic veterans of 45, 48–9
  civil defence in 75, 77–8, 129
  compensation by 48
  foreign policy under Trump 132–5, 136
  inequality in 24–5
  near misses and false alarms 97–101
  pacifism in 107–8, 111–13, 114, 115–18
  scientific workers in 16–20, 107–8
  secret cities in 2, 3
  testing by 30–5, 54, 55, 58–60, 113
  see also Los Alamos

unmedicalised conditions 46
uranium and plutonium storage 102, 103
uranium mines 7, 64, 70
US-UK Mutual Defence Agreement 33
USSR *see* Soviet Union
Uyghur (Uighur) community 65–6
Uyoku dantai 28–9
Uzbekistan 127

vertical proliferation 91

waste repositories 54, 60
Western Samoa 60
Wisner, B. 81
women
  local, medical examinations of 59
  as pacifists 106, 109–10, 120
  as politicians 5–6
  as scientific workers 18–20
Women Strike for Peace 109
Woodward, R. 71
World War II 10–16 *see also* Hiroshima; Los Alamos; Nagasaki
WPC (World Peace Council) 114, 118–19
Wright, Tim 122–3

'x-ray hands', myth of 47
Xianjiang province 65–6

Y-12 National Security Complex 10
Yeltsin, Boris 99

Zamyatin, Yevgeny 2

The Pluto Press Newsletter

Hello friend of Pluto!

Want to stay on top of the best radical books we publish?

Then sign up to be the first to hear about our new books, as well as special events, podcasts and videos.

You'll also get 50% off your first order with us when you sign up.

Come and join us!

Go to bit.ly/PlutoNewsletter